METHODS IN MOLECULAR BIOLOGY

Series Editor
John M. Walker
School of Life and Medical Sciences
University of Hertfordshire
Hatfield, Hertfordshire, AL10 9AB, UK

For further volumes:
http://www.springer.com/series/7651

Traumatic and Ischemic Injury

Methods and Protocols

Edited by

Binu Tharakan

Department of Surgery, Texas A&M University Health Science Center, College of Medicine,
Baylor Scott and White Research Institute, Temple, TX, USA

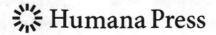 Humana Press

Editor
Binu Tharakan
Department of Surgery
Texas A&M University Health Science Center
College of Medicine, Baylor Scott
and White Research Institute
Temple, TX, USA

ISSN 1064-3745 ISSN 1940-6029 (electronic)
Methods in Molecular Biology
ISBN 978-1-4939-8510-4 ISBN 978-1-4939-7526-6 (eBook)
https://doi.org/10.1007/978-1-4939-7526-6

Cover illustration: Endothelial cells of the blood-brain barrier stained for zonula occludents-1 (green) and nucleus (blue).

Printed on acid-free paper

This Humana Press imprint is published by the registered company Springer Science+Business Media, LLC part of Springer Nature.
The registered company address is: 233 Spring Street, New York, NY 10013, U.S.A.

Preface

Traumatic and ischemic injuries are some of the most devastating diseases and are responsible for high mortality and morbidity worldwide. Over the past decades, significant research has been conducted in this area, aimed at reducing or preventing complications after a traumatic or ischemic injury. Although some advances have been made in our understanding of the pathobiology of such insults, further research and innovative approaches are essential to clearly understand and treat these complex issues.

The use of a suitable animal model that exhibits many of the pathophysiological features of any disease is important for the identification of therapeutic targets and future drug development to alleviate its symptoms. A wide variety of small and large animal models are currently available for researchers working in the field of traumatic and ischemic injuries. The major advantage of using animals in research in these fields is the ability to obtain scientific information under well-defined conditions that closely mimic specific clinical scenarios. So, in the field of traumatic and ischemic injuries, it is important to continue animal studies and to develop new animal models and protocols that better reflect the clinical setting to gain further insight into the underlying cellular, biochemical, and physiological mechanisms involved. The general trend in the field is to use laboratory rodents to answer fundamental questions regarding cellular and molecular mechanisms while physiological processes are more commonly investigated in large animals due to the similarity to humans. The main focus of this volume was to compile the procedures for the development and application of several of such research models and techniques with the help from experts in their respective fields.

This particular volume is part of the *Methods in Molecular Biology* series and I hope will allow beginners with limited experience to initiate new research projects and established investigators to identify other comparable approaches and strategies. The first few chapters are mainly focused on some of the selected animal models and procedures commonly used in traumatic injury research followed by a chapter on hemorrhagic stroke. After that, a series of chapters on various ischemic injuries followed by some of the chapters specifically addressing sepsis models have been included. The last few chapters are dedicated to selected in vitro models from the field.

I would like to thank everyone who helped me directly or indirectly with this project. I am very much indebted to all the contributing authors for their time, hard work, and support for this project. I would like to express my gratitude to Ms. Chinchusha Anasooya Shaji and Dr. Himakarnika Alluri for their help in the organizing and editing processes. Also, my deepest appreciation goes to Prof. John Walker, the series editor, for his constant encouragement, guidance, and understanding, without which this project would not have been successful at all.

Temple, TX, USA   *Binu Tharakan*

Contents

Contributors

HIMA C.S. ABEYSINGHE • *Department of Medicine, University of Melbourne, St Vincent's Campus, Fitzroy, VIC, Australia*

ONAT AKYOL • *Department of Physiology & Pharmacology, Loma Linda University, Loma Linda, CA, USA*

HIMAKARNIKA ALLURI • *Department of Surgery, Texas A&M University Health Science Center, College of Medicine, Baylor Scott and White Research Institute, Temple, TX, USA*

NATASCHA G. ALVES • *Department of Molecular Pharmacology and Physiology, Morsani College of Medicine, University of South Florida, Tampa, FL, USA*

ERIC X. BECK • *Department of Physiology and Cell Biology, Davis Heart and Lung Research Institute, Ohio State University Wexner Medical Center, Columbus, OH, USA*

JEROME W. BRESLIN • *Department of Molecular Pharmacology and Physiology, Morsani College of Medicine, University of South Florida, Tampa, FL, USA*

SHERREFA BURCHELL • *Department of Physiology & Pharmacology, Loma Linda University, Loma Linda, CA, USA*

MALGORZATA BUREK • *Department of Anaesthesia and Critical Care, University of Wuerzburg, Wuerzburg, Germany*

ED W. CHILDS • *Department of Surgery, Morehouse School of Medicine, Atlanta, GA, USA*

SOUMEN CHOUDHURY • *Department of Pharmacology and Toxicology, College of Veterinary Science & Animal Husbandry, U.P. Pandit Deen Dayal Upadhyaya Pashu Chikitsa Vigyan Vishwavidyalaya Evam Go-Anusandhan Sansthan, Mathura, Uttar Pradesh, India*

FREDERICK COLBOURNE • *Department of Psychology, Neuroscience and Mental Health Institute, University of Alberta, Edmonton, AB, Canada*

ALAN DARDIK • *Yale University School of Medicine, New Haven, CT, USA*

MATTHEW L. DAVIS • *Department of Surgery, Texas A&M University Health Science Center, College of Medicine, Baylor Scott and White Research Institute, Temple, TX, USA*

TRAVIS M. DOGGETT • *Department of Molecular Pharmacology and Physiology, Morsani College of Medicine, University of South Florida, Tampa, FL, USA*

PHILIP A. EFRON • *Department of Surgery, Shands Hospital, University of Florida College of Medicine, Gainesville, FL, USA*

PRINCE ESIOBU • *Department of Surgery, Morehouse School of Medicine, Atlanta, GA, USA*

ROGER G. EVANS • *Department of Physiology, Monash University, Parkville, VIC, Australia*

CAROLA Y. FÖRSTER • *Department of Anaesthesia and Critical Care, University of Wuerzburg, Wuerzburg, Germany*

LORETTA INIAGHE • *Department of Physiology & Pharmacology, Loma Linda University, Loma Linda, CA, USA*

ROSELEEN F. JOHN • *Neuroscience and Mental Health Institute, University of Alberta, Edmonton, AB, Canada*

JASON A. JUSTICE • *Department of Surgery, Texas A&M University Health Science Center, College of Medicine, Temple, TX, USA*

PAIGE S. KATZ • *Department of Physiology, School of Medicine, Louisiana State University Health Science Center, New Orleans, LA, USA*

DAMON KLEBE • *Department of Physiology & Pharmacology, Loma Linda University, Loma Linda, CA, USA*

JUNKO KOSAKA • *Florey Institute of Neuroscience and Mental Health, University of Melbourne, Parkville, VIC, Australia*

YUGEESH R. LANKADEVA • *Florey Institute of Neuroscience and Mental Health, University of Melbourne, Parkville, VIC, Australia*

TYLER J. LOFTUS • *Department of Surgery, University of Florida College of Medicine, Gainesville, FL, USA*

ANGELA LOMAS • *Department of Surgery, Texas A&M University Health Science Center, College of Medicine, Baylor Scott and White Research Institute, Temple, TX, USA*

MARY SUSAN LOPEZ • *Department of Neurosurgery, University of Wisconsin-Madison, Madison, WI, USA*

ROBERT T. MALLET • *Department of Integrative Physiology and Anatomy, University of North Texas Health Science Center, Fort Worth, TX, USA*

BRITTANY MATHIAS • *Department of Surgery, University of Florida College of Medicine, Gainesville, FL, USA*

CLIVE N. MAY • *Florey Institute of Neuroscience and Mental Health, University of Melbourne, Parkville, VIC, Australia*

KEVIN E. MCELHANON • *Department of Physiology and Cell Biology, Davis Heart and Lung Research Institute, Ohio State University Wexner Medical Center, Columbus, OH, USA*

JUAN C. MIRA • *Department of Surgery, University of Florida College of Medicine, Gainesville, FL, USA*

SANTOSH KUMAR MISHRA • *Division of Pharmacology & Toxicology, Indian Veterinary Research Institute, Bareilly, Uttar Pradesh, India; Bhubaneswar, Odisha, India*

ALICIA M. MOHR • *Department of Surgery, University of Florida College of Medicine, Gainesville, FL, USA*

LYLE L. MOLDAWER • *Department of Surgery, University of Florida College of Medicine, Gainesville, FL, USA*

PATRICIA E. MOLINA • *Department of Physiology, School of Medicine, Louisiana State University Health Science Center, New Orleans, LA, USA*

DINA C. NACIONALES • *Department of Surgery, University of Florida College of Medicine, Gainesville, FL, USA*

CESAR REIS • *Department of Physiology & Pharmacology, Loma Linda University, Loma Linda, CA, USA*

BOBBY DARNELL ROBINSON • *Department of Surgery, Texas A&M University Health Science Center, College of Medicine, Baylor Scott and White Research Institute, Temple, TX, USA*

CARLI L. ROULSTON • *Department of Medicine, University of Melbourne, St Vincent's Campus, Fitzroy, VIC, Australia*

MYOUNG-GWI RYOU • *Department of Medical Laboratory Sciences and Public Health, Tarleton State University, Fort Worth, TX, USA; Department of Integrative Physiology and Anatomy, University of North Texas Health Science Center, Fort Worth, TX, USA*

ELLAINE SALVADOR • *Department of Anaesthesia and Critical Care, University of Wuerzburg, Wuerzburg, Germany*

RUSSELL M. SANCHEZ • *Division of Neurology, Department of Pediatrics, Emory University School of Medicine, Atlanta, GA, USA*

CHINCHUSHA ANASOOYA SHAJI • *Department of Surgery, Texas A&M University Health Science Center, College of Medicine, Baylor Scott and White Research Institute, Temple, TX, USA*

JIPING TANG • *Department of Physiology & Pharmacology, Loma Linda University, Loma Linda, CA, USA*

BINU THARAKAN • *Department of Surgery, Texas A&M University Health Science Center, College of Medicine, Baylor Scott and White Research Institute, Temple, TX, USA*

SRINIVAS M. TIPPARAJU • *Department of Molecular Pharmacology and Physiology, Morsani College of Medicine, University of South Florida, Tampa, FL, USA; Department of Pharmaceutical Sciences, College of Pharmacy, University of South Florida, Tampa, FL, USA*

JARED J. TUR • *Department of Pharmaceutical Sciences, College of Pharmacy, University of South Florida, Tampa, FL, USA*

RICARDO UNGARO • *Department of Surgery, University of Florida College of Medicine, Gainesville, FL, USA*

RAGHU VEMUGANTI • *Department of Neurosurgery, University of Wisconsin-Madison, Madison, WI, USA*

WENDY E. WALKER • *Center of Emphasis in Infectious Diseases, Department of Biomedical Sciences, Texas Tech University Health Sciences Center El Paso, El Paso, TX, USA*

NOAH WEISLEDER • *Department of Physiology and Cell Biology, Davis Heart and Lung Research Institute, Ohio State University Wexner Medical Center, Columbus, OH, USA*

KATIE WIGGINS-DOHLVIK • *Department of Surgery, Texas A&M University Health Science Center, College of Medicine, Baylor Scott and White Research Institute, Temple, TX, USA*

MICHAEL R. WILLIAMSON • *Neuroscience and Mental Health Institute, University of Alberta, Edmonton, AB, Canada*

ZHAOBIN XU • *Department of Physiology and Cell Biology, Davis Heart and Lung Research Institute, Ohio State University Wexner Medical Center, Columbus, OH, USA*

JUN YU • *Yale University School of Medicine, New Haven, CT, USA*

SARAH Y. YUAN • *Department of Molecular Pharmacology and Physiology, Morsani College of Medicine, University of South Florida, Tampa, FL, USA*

JOHN H. ZHANG • *Department of Physiology & Pharmacology, Loma Linda University, Loma Linda, CA, USA*

Chapter 1

Mouse Injury Model of Polytrauma and Shock

Juan C. Mira, Dina C. Nacionales, Tyler J. Loftus, Ricardo Ungaro, Brittany Mathias, Alicia M. Mohr, Lyle L. Moldawer, and Philip A. Efron

Abstract

Severe injury and shock remain major sources of morbidity and mortality worldwide. Immunologic dysregulation following trauma contributes to these poor outcomes. Few, if any, therapeutic interventions have benefited these patients, and this is due to our limited understanding of the host response to injury and shock. The Food and Drug Administration requires preclinical animal studies prior to any interventional trials in humans; thus, animal models of injury and shock will remain the mainstay for trauma research. However, adequate animal models that reflect the severe response to trauma in both the acute and subacute phases have been limited. Here we describe a novel murine model of polytrauma and shock that combines hemorrhagic shock, cecectomy, long bone fracture, and soft-tissue damage. This model produces an equivalent Injury Severity Score associated with adverse outcomes in humans, and may better recapitulate the human leukocyte, cytokine, transcriptomic, and overall inflammatory response following injury and hemorrhagic shock.

Key words Mouse, Polytrauma, Hemorrhage, Inflammation, Immunity

---

1 Introduction

Trauma remains one of the leading causes of death in all age groups [1, 2]. Implementation of timely and standardized resuscitative interventions, as well as advances in critical care medicine, have improved early survival in trauma. However, the morbidity and mortality from late complications after severe injury remains high [2–5]. In an effort to identify the etiology and immunologic basis for late multiple organ failure (MOF) that contributes to death after severe trauma, a number of paradigms have been established and revised over the past three decades [5–8]. Chronic critical illness, or CCI, and the persistent inflammation, immunosuppression, and protein catabolism syndrome or PICS, now define a population of

Electronic supplementary material: The online version of this chapter (https://doi.org/10.1007/978-1-4939-7526-6_1) contains supplementary material, which is available to authorized users.

Binu Tharakan (ed.), *Traumatic and Ischemic Injury: Methods and Protocols*, Methods in Molecular Biology, vol. 1717, https://doi.org/10.1007/978-1-4939-7526-6_1, © Springer Science+Business Media, LLC 2018

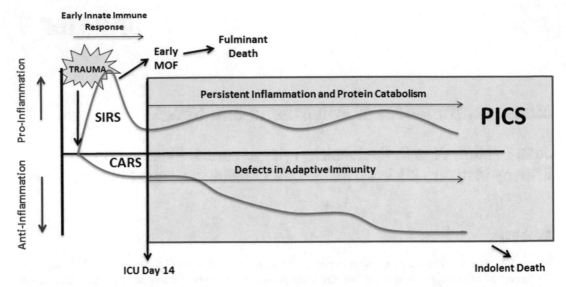

Fig. 1 Persistent inflammation-immunosuppression catabolism syndrome (PICS). Improved methods of trauma resuscitation have created a population of intensive care unit (ICU) patients who survive their initial insult and early multiple organ failure (MOF) to remain in the ICU for prolonged periods of time with manageable organ dysfunction, malnourishment despite nutritional intervention, suffer from recurrent infections, and rarely rehabilitate to a functional life. Adapted from [5]

ICU patients that survive their initial insult, but remain in the ICU for prolonged periods of time [5, 9] (*see* Fig. 1). These patients have manageable organ dysfunction, are malnourished despite nutritional intervention, suffer from recurrent infections, and rarely rehabilitate to a functional life [5, 10].

There have been numerous attempts at therapeutic interventions meant to improve early survival and prevent the development of PICS after trauma [11, 12]. Unfortunately, efforts at pharmacologic modulation of the immunologic and pathophysiologic response to injury (or infection) in humans have failed, despite promising preclinical results in murine models [11, 12]. This failure to repeat the success of preclinical murine studies in humans has raised questions about the suitability of current murine models of severe inflammation [13–16]. In particular, critics have highlighted the genetic homogeneity of the inbred mouse, the genome-wide differences between man and mice, the heterogeneity of the human condition, as well as the specific variance in cellular composition between murine and human tissues [13, 15]. Thus, researchers are still attempting to improve our methodology for investigating the consequences of severe shock and injury.

Numerous animal models of trauma and shock have been employed over the past century [17–20]. One of the earliest described murine models of trauma was the Noble–Collip drum, described in 1942. In this model, a mouse without anesthesia is tumbled repeatedly in a metal drum, leading to intra-abdominal

Table 1
Examples of murine models of hemorrhagic shock, trauma, burn, or combination

Model name	Modifications to make more similar to human condition
Trauma-hemorrhage (T-H) model followed by cecal ligation and puncture	Two hit model of hemorrhagic shock followed by polymicrobial sepsis, usually 24 h later [46, 47]
T-H model with chronic restraint	Hemorrhagic shock combined with pulmonary contusion and subsequent chronic restraint and stress of the rodent [48]
Pseudofracture with or without T-H	Recreates the features of long bones without breaking the native bone. A bilateral muscle crush injury to the hind limbs is performed, followed by injection of a bone solution into these injured muscles [49]
T-H with fracture and associated soft tissue injury	Hemorrhagic shock combined with a closed mid-shaft fracture of the femur and fracture related soft tissue injury [21]
T-H with fracture and midline laparotomy	Hemorrhagic shock combined with a closed mid-shaft fracture of the femur and midline laparotomy to reflect soft tissue injury [23, 24]
Traumatic brain injury (TBI)	Clinically relevant murine model using an impacting rod directly on the skull [50]
Burn injury followed by infection	Clinically relevant severe burn model that subsequently subjects the mouse to a secondary infection such as pneumonia. This attempts to recreate secondary infections and morbidity and mortality that are common in human burn patients [51]

organ injury, diffuse bruising with muscle injury, and death [20]. Subsequently, researchers developed individual models of hemorrhagic shock, thoracic trauma, brain injury, long bone fracture with soft tissue injury, and intra-abdominal trauma [18, 19, 21–27]. Some of these models have been combined in an attempt to better imitate the human condition (*see* Table 1). These previous rodent models of trauma and shock have made significant contributions to our understanding of the biological response to injury, yet critical analysis and revision of these models are necessary with technological advances and our expanding knowledge of the biology of inflammation, particularly at the genomic level [15, 16, 18, 28–30].

The criticisms surrounding current murine trauma and shock models are longstanding [15–19, 28, 29], but recent concerns stem from studies that have further highlighted the inherent differences in the murine and human genomic response to inflammatory diseases [13, 14, 16, 31]. The Mouse ENCODE Consortium [32] and the "Inflammation and Host Response to Injury" (Glue Grant) [30, 33] have catalogued the transcriptomic response in both health and disease in mice and humans. These human and murine datasets have been extensively evaluated by different groups, often leading to contradictory conclusions on whether the genomic response to inflammatory disease in mice recapitulates the human response to severe injury of infection [13, 14]. Although there is considerable controversy over the value of these murine models, there is general consensus that these animals models have some value but must be used with caution and an appropriate understanding of their limitations—this is the only way translatable progress in immunotherapy will occur in the future [34–36] .

One fundamental criticism of existing murine models is that they fail to recapitulate the severity and multicompartmental nature of human trauma [16, 30, 31, 37]. Also, animal welfare issues appropriately limit our ability to completely imitate human trauma [38]. However, our laboratory has developed a murine model that better reflects the early inflammatory response of the severely injured patient while remaining within the guidelines of humane treatment of laboratory animals [31, 39, 40]. In an effort to better replicate human trauma in a murine model, we focused on combining insults that produce an Injury Severity Score greater than 15 [26]. The Injury Severity Score (ISS) is an anatomical scoring system that provides an overall score for patients with multiple injuries [41]. This value is intended to accurately represent the patient's degree of critical illness. It has been considered the gold standard of classifying trauma patients early after their injury: Minor (ISS 1–3), Moderate (ISS 4–8), Serious (ISS 9–15), Severe (ISS 16–24), and Critical (ISS 25–75). An ISS greater than 15 is generally used as the minimum score for trauma studies of severely injured patients, as these individuals are more likely to have poor outcomes [1, 39, 41] .

The polytrauma model described in this text utilizes the well-established trauma hemorrhage murine model of 90 min of hemorrhagic shock [42, 43]. In addition, a cecectomy, a long bone fracture, and muscle tissue damage are subsequently carried out in the mouse under anesthesia. The combination of these elements creates an insult which equates to an ISS of 18 in a human [39]. Although a further increased ISS would likely lead to a superior model that better represents human trauma, Institutional Animal Care and Use Committee's (IACUC) limitations that rightly guard the humane treatment of animals typically will not allow for further insult to the rodents. Notwithstanding, in our experience, we have found that this model better reproduces the human immune response to injury and shock [31, 39].

Mice subjected to the polytrauma model below have an inflammatory response that better reflects the cytokine, chemokine, and leukocyte reaction seen after human trauma [39]. For example, we demonstrated that the plasma concentrations of interleukin-10 (IL-10), interferon-inducible protein-10 (IP-10), macrophage

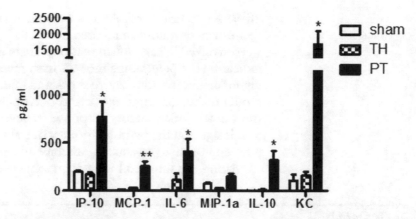

Fig. 2 Plasma cytokine concentration in sham, trauma-hemorrhage (TH), and polytrauma (PT) mice. *$p < 0.05$, **$p < 0.01$ vs sham. Adapted from [39]

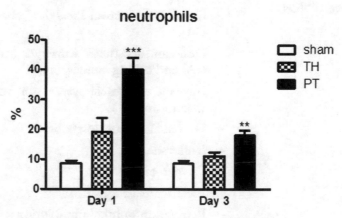

Fig. 3 Blood neutrophils in sham, trauma-hemorrhage (TH), and polytrauma (PT) mice. **$p < 0.01$, ***$p < 0.001$ vs sham. Adapted from [39]

inflammatory protein-1α (MIP-1α), keratinocyte chemoattractant (KC), interleukin-6 (IL-6), and monocyte chemoattractant portein-1 (MCP-1) were significantly greater when compared to a traditional trauma hemorrhage model and more comparable to what is displayed in humans after trauma [39] (*see* Fig. 2). Furthermore, these animals had a leukocytosis associated with neutrophilia (*see* Fig. 3), had decreased major histocompatibility complex class II expression in bone marrow myeloid cells, and both of these effects were sustained beyond 1 day after the operation in this model; these phenomena were not displayed in the historical murine models of hemorrhagic shock and minor trauma [39]. In addition, this new model demonstrated improved correlations between the leukocyte gene expression patterns of severe human trauma patients, although species differences were still very evident [31]. However, when we evaluated the top responsive genes to trauma in humans,

88–99% of murine orthologues were identified to respond similarly to human transcriptomic alterations [31]. Many of these genes were involved in early inflammation, innate and adaptive immunity, indicating the polytrauma model's usefulness as a method of studying inflammation after severe shock and injury [31]. We believe this model of hemorrhagic shock and polytrauma better recapitulates the human inflammatory response to severe injury and provides new insights in the pathophysiology of trauma while also offering a new approach to evaluating therapeutic interventions aimed at alleviating the short and long-term sequela of traumatic injury.

2   Materials

2.1 Materials for Anesthesia and Preparation

– Sterile gloves.
– Weighing scale.
– One-liter induction Plexiglas® chamber (VetEquip, Livermore, CA).
– Desiccant anesthetic scavenger canister (VetEquip, Livermore, CA) and tubing connectors.
– Inhalant anesthesia system for veterinary surgery (VetEquip, Livermore, CA).
– Universal rodent nosecone.
– Isoflurane.
– Oxygen tank.
– Electric razor.
– Petrolatum ophthalmic solution.
– Platform for animal surgery (6 × 6 in. Plexiglas® plates).
– Fisher™ paper tape, 0.5 in.
– Vis-U-All Self Seal sterilization pouch (Weck, A Squibb Company; Princeton, NJ).

2.2 Materials for Surgery

– Mice (*see* **Note 1**).
– Blood pressure analyzer (Micro-Med, Louisville, KY).
– Dissecting microscope (Scienscope, Matthews, NC).
– Bead sterilizer.
– Infusion pump.
– Polyethylene (PE)-10 catheter tubing (*see* **Note 2**).
– Betadine scrubs.
– Sterile water.
– Lidocaine.

- Heparin.
- Normal saline.
- Lactated Ringer's (LR) solution.
- Microforceps, angled, 3.5 in, 0.3 × 5 mm.
- Nontoothed forceps.
- Microdissection scissors, 4.5 in.
- Microvessel clips, 0.75 × 4 mm.
- Bulldog vessel clips.
- Small scissors.
- Small needle driver.
- Small bone cutter.
- 30 g, ½ in., hypodermic needles.
- 1 mL syringe—slip tip.
- 5-0 silk suture.
- 2-0 silk suture.
- 5-0 Ethilon suture.
- 3-0 Vicryl suture.
- BD Autoclip Closing System™.
- Auto-clips, 7 mm.
- Buprenorphine for injection (*see* **Note 3**).
- Circulating water warming pad.

3 Methods

3.1 Induction and preparation

1. All instruments must be sterilized either by heat, ethylene oxide or other approved method. Merely cleaning, sanitizing or using antimicrobial washes are not adequate. In addition, all surgical materials such as sutures or sponges must be sterile and still within their expiration date. Solutions used for animals such as normal saline, Lactated Ringers (LR) solution or other drugs must be sterile and either the US Pharmacopeial Convention (USP) compliance standard or veterinary-grade (unless otherwise approved by the local IACUC).

2. Weigh the mouse (*see* **Note 4**).

3. Place the mouse in the induction chamber. Adjust the isoflurane concentration to 3.5–4.5% with O_2 flow at 1 L/min until an adequate plane of anesthesia is reached (*see* **Note 5**).

4. Shave the abdomen and inguinal area.

5. Apply USP grade petrolatum ophthalmic ointment to both eyes to prevent drying during the procedure.

6. Place the mouse on the Plexiglas® plate or other nonabsorbent board in the supine position.

7. Place the anesthesia cone to the snout of mouse and maintain isoflurane concentration at 1.5–2.0% with 100% O_2 flow at 0.5–1 L/min.

8. Restrain the mouse by taping all four extremities and its tail (*see* **Note 6**).

3.2 Femoral Artery Catheterization

1. Wipe the inguinal area with Betadine™ scrub and sterile water, alternating three times.

2. Place the mouse on a water-circulating, nonabsorbent warming pad under a dissecting microscope.

3. Visualize the femoral vessels through the skin, lift the overlying skin with forceps and excise a small section of skin to expose the femoral vessels using the microdissection scissors.

4. Carefully separate the femoral nerve and vein from the artery by blunt dissection using the microforceps.

5. Pass two 5-0 silk ties under the femoral artery spaced approximately 0.5 cm apart.

6. Ligate the femoral artery with the distal tie (5-0 silk) and attach a bulldog clip to the loose ends of the distal tie (*see* **Note 7**).

7. Tie the proximal tie (5-0 silk) loosely by creating a knot but not tightening it in order to keep the femoral artery lumen patent.

8. Place a vessel clip proximal to the proximal tie.

9. Apply a drop of 1% lidocaine to area (*see* **Notes 8** and **9**).

10. Puncture the artery just above the distal tie using a 30-gauge needle. The needle should have its bevel up such that the whole in the needle is facing the ventral or anterior portion of the mouse.

11. Gently insert the PE-10 catheter attached to a 1 mL syringe filled with LR solution, again with the bevel facing up (*see* **Notes 10** and **11**).

12. Remove the vessel clip from the femoral artery.

13. Gently advance the catheter using microforceps (*see* Image 1).

14. Secure the catheter by the proximal tie (5-0 silk) by tightening the knot made in **step 7**.

15. Another 5-0 silk tie may be added and tied to secure the catheter.

16. Tape the catheter to the Plexiglas® plate to further secure it in place.

17. Repeat **steps 3** through **16** on the contralateral side.

3.3 Hemorrhagic Shock and Resuscitation

1. Remove the mouse from the inhalational anesthesia and remove the 1 mL syringe from one of the catheters and attach this catheter to the pressure transducer line (*see* **Notes 12** and **13**).

2. Allow the mouse to emerge from anesthesia (at this point its mean arterial pressure (MAP) should be ~95 mmHg).

3. Reattach a new 1 mL syringe flushed with 1000 USP units/mL heparin to the contralateral catheter and slowly aspirate blood until the animal's MAP is 30–40 mmHg (*see* Image 2).

4. Maintain the mouse in hemorrhagic shock for 90 min (*see* **Notes 9** and **14**).

5. Attach the catheter used for hemorrhagic shock to the infusion pump.

6. Resuscitate the mouse by administrating of LR solution. The volume of LR will be four times the volume of blood removed, and the infusion rate of the LR solution should be at a rate of 10 mL/h.

3.4 Decannulation and Bone Fracture

1. Disconnect the catheter connected to the blood pressure transducer and attach a 1 mL syringe filled with LR. Also disconnect the catheter connected to the infusion pump and attach a new 1 mL syringe filled with LR.

2. Reanesthetize the mouse. Place the anesthesia cone to the snout of mouse and maintain isoflurane concentration at 1.5–2.0% with O_2 flow at 0.5 L/min. Again, ensure an adequate plane of anesthesia is reached (*see* **Note 5**).

3. Gently pull back the catheter until just past the proximal tie.

4. Ligate the artery by further tightening the proximal tie.

5. Completely remove the catheter and discard it in appropriate trash bag/container.

6. On one side, proceed to bluntly dissect the adjacent muscles to expose the long bone (tibia).

7. Fracture the bone using a bone cutter or small scissors (*see* Image 3).

8. Realign the bone (*see* Image 4).

9. Grasp the superior muscle tissue with a clamp for 30 s

10. Close the inguinal incisions with a running stitch or with a figure of eight stitch using a 5-0 Ethilon suture.

3.5 Cecectomy

1. Wipe the abdomen with Betadine™ scrub and sterile water, alternating the solutions three times.

2. Apply the sterile drape.

3. Make a 1 cm midline skin incision.

    (a) Using either straight-edge or iris scissors, make a 1 cm midline cut into the skin only, approximately 0.5–1 cm away from xiphoid process.

    (b) Entry into the peritoneum should be avoided.

4. Identify the abdominal wall and cut through the abdominal musculature and into the peritoneum (*see* **Notes 15** and **16**).

5. Identify and exteriorize the cecum using nontoothed forceps.

6. Doubly clamp the cecum about 1 cm from the tip (*see* **Note 17**) (*see* Image 5).

7. Ligate the stump with silk 2-0 using at least one square knot.

8. Excise the cecum (*see* Image 6).

9. Return the cecal stump to the abdominal cavity.

10. Close the abdominal wall using an absorbable suture (3-0 Vicryl) using a running suture.

11. Close the skin using surgical clips.

12. Inject the mouse subcutaneously with 0.05–1 mg/kg buprenorphine in 1 mL saline (*see* **Note 18**).

13. Return the animals to their cages after they awaken from anesthesia. This is usually indicated by their capacity to independently turn over onto their bellies after being placed on their sides or back. Keep the mouse on the warming pad until fully recovered.

14. Continue to monitor the mouse until it has fully recovered from anesthesia (*see* **Note 19**).

15. If additional mice are to be used, the instrument tips are to be washed in distilled water taking care to remove any blood or tissue. The instruments are then placed into a hot bead sterilizer for at least 10 min while the next animal is prepared. The sterile field should not be broken and the instruments are not to be handled without sterile gloves.

4 Notes

1. The preoperative care of mice used in research models should not be overlooked. In our experience, allowing the mice to equilibrate to the diurnal cycle and feeding at the research institution for 1–2 weeks before manipulations decreases model variability.

2. The catheters are prepared by cutting the PE-10 tubing into to 6–8 in. catheters. Then a 27-gauge needle is filed down and inserted into one end of the tubing to allow for the connection to 1 mL syringe or blood pressure transducer as later described.

The catheters are gas sterilized in Vis-U-All Self Seal sterilization pouches.

3. Buprenorphine is a controlled substance and will require appropriate storage and documentation. It is given subcutaneously in 1 mL saline at a dose of 0.05–0.2 mg/kg body weight.

4. The *Guide* (*see* ref. 44) requires that scales used to weigh mice for weight-based injections be calibrated at least twice annually.

5. Observe the mice for the proper anesthetic depth. An adequate plane of anesthesia is monitored by assessing muscle tone and the absence of reflex to surgical manipulation, toe pinch and ocular reflex.

6. Taping is critical in keeping the mouse restrained, especially during the hemorrhagic shock portion of the model.

7. The bulldog clip will provide countertraction during cannulation.

8. Lidocaine is used to allow for dilation of the artery which facilitates cannulation.

9. Make sure to keep the field moist with 1% lidocaine or LR solution. If the field dries the vessels are likely to rupture and there will be poor blood flow through the catheter. Continue to monitor the animal during the entire period of shock (90 min) and keep field moist as necessary.

10. The catheter tip should be modified to match the diameter of the artery. If the catheter is too thick it will not enter the vessel. If the catheter is too thin, blood will clot within the catheter. To modify the tip, gently pull and stretch the catheter tip until it matches the diameter of the femoral artery.

11. The catheter tip is cut at an angle of about 30° to form a bevel point. If the angle is too steep it is likely to puncture the back wall. If it is too shallow it will not enter the vessel.

12. Place the mouse on a water-circulating heating pad during hemorrhagic shock and resuscitation period. Do not use an electric heating pad or a lamp, as the risk of skin burns to the rodent is increased.

13. The catheter will be attached to the pressure transducer for the full length of the period of hemorrhagic shock to provide continuous blood pressure measurements.

14. Maintain the MAP at 30–40 mmHg by repeat blood draws or LR solution infusion. Continue to monitor the animal during the entire period of shock. This period of shock while not being under anesthesia is an important component to inducing the appropriate immune and pathophysiologic response in the mouse. Also, ensure normothermia throughout the entire hemorrhage and resuscitation periods.

15. Generally, the cecum will be located directly under the midline incision and will be easy to exteriorize.

16. The small bowel can be easily transected during entry into the peritoneum. Therefore, care needs to be taken not to inadvertently damage the bowel with the scissors as you enter the abdominal cavity. Furthermore, excessive manipulation of the intestines will result in bleeding, perforation, and/or serosal tears as it is very fragile.

17. It is very important that the ligation be made distal to the area where the ileum enters the cecum (ileocecal junction) to prevent obstruction of the mouse's intestines.

18. The use of opioid analgesics is strongly recommended in this model prior to arousal from anesthesia and every 12 h until euthanasia to ameliorate pain and distress associated with surgical injury, soft tissue injury and long bone fracture.

19. Monitoring procedures are used to determine when and if the mice are to enter a prodromal period where death is likely to occur. The mice should not be allowed to die spontaneously, and instead be euthanized when they become moribund and death is imminent for well-established ethical and scientific reasons. This will reduce pain and discomfort to the animal and allows samples to be collected and not lost secondary to postmortem changes. Animals should be monitored for changes in posture, failure to right themselves, and signs of distress as previously described [45].

    (a) Changes in posture—soon after the procedure the mice will appear hunched; however, this will resolve with time. If the mice fail to thrive, the following may be considered: moist chow, more resuscitation with normal saline, or euthanasia depending on your protocol.

    (b) Failure to right themselves—a mouse that is placed on their back should be able to right themselves regardless of femur fractures and muscle injury. Mice that are unable to do so must be euthanized.

    (c) Signs of distress including but not limited to pain on touch and, agonal breathing—In these situations, further resuscitation may be considered or progression to euthanasia of the mouse depending on your protocol.

Acknowledgment

J.C.M., T.L.M., and B.M. were supported by a training grant in burn and trauma research from the National Institute of General Medical Sciences (NIGMS) (T32 GM-008721). This work was also supported by NIH Grants R01 GM-040586 and R01 GM-081923, awarded by the NIGMS. A.M.M. was supported by R01 GM-105893. In addition, P.A.E. was supported by P30 AG-028740 from the NIH National Institute on Aging and R01 GM113945 (NIGMS). Finally, P.A.E. and L.L.M. were supported by P50 GM-111152 (NIGMS).

References

1. Pfeifer R, Tarkin IS, Rocos B, Pape HC (2009) Patterns of mortality and causes of death in polytrauma patients—has anything changed? Injury 40:907–911

2. Probst C, Pape HC, Hildebrand F, Regel G, Mahlke L, Giannoudis P, Krettek C, Grotz MR (2009) 30 years of polytrauma care: an analysis of the change in strategies and results of 4849 cases treated at a single institution. Injury 40:77–83

3. Probst C, Zelle BA, Sittaro NA, Lohse R, Krettek C, Pape HC (2009) Late death after multiple severe trauma: when does it occur and what are the causes? J Trauma 66 (4):1212–1217

4. Sasser SM, Varghese M, Joshipura M, Kellermann A (2006) Preventing death and disability through the timely provision of prehospital trauma care. Bull World Health Organ 84:507

5. Gentile LF, Cuenca AG, Efron PA, Ang D, Bihorac A, McKinley BA, Moldawer LL, Moore FA (2012) Persistent inflammation and immunosuppression: a common syndrome and new horizon for surgical intensive care. J Trauma Acute Care Surg 72:1491–1501

6. Robertson CM, Coopersmith CM (2006) The systemic inflammatory response syndrome. Microbes Infect 8:1382–1389

7. Ward NS, Casserly B, Ayala A (2008) The compensatory anti-inflammatory response syndrome (CARS) in critically ill patients. Clin Chest Med 29:617–625. viii

8. Rosenthal MD, Moore FA (2015) Persistent inflammatory, immunosuppressed, catabolic syndrome (PICS): a new phenotype of multiple organ failure. J Adv Nutr Hum Metab 1:e874

9. Mira JC, Gentile LF, Mathias BJ, Efron PA, Brakenridge SC, Mohr AM, Moore FA, Moldawer LL (2017) Sepsis pathophysiology, chronic critical illness, and persistent inflammation-immunosuppression and catabolism syndrome. Crit Care Med 45:253–262

10. Mathias B, Delmas AL, Ozrazgat-Baslanti T, Vanzant EL, Szpila BE, Mohr AM, Moore FA, Brakenridge SC, Brumback BA, Moldawer LL et al (2016) Human myeloid-derived suppressor cells are associated with chronic immune suppression after severe sepsis/septic shock. Ann Surg 265:827–834

11. Stahel PF, Smith WR, Moore EE (2007) Role of biological modifiers regulating the immune response after trauma. Injury 38:1409–1422

12. Artenstein AW, Higgins TL, Opal SM (2013) Sepsis and scientific revolutions. Crit Care Med 41:2770–2772

13. Seok J, Warren HS, Cuenca AG, Mindrinos MN, Baker HV, Xu W, Richards DR, McDonald-Smith GP, Gao H, Hennessy L et al (2013) Genomic responses in mouse models poorly mimic human inflammatory diseases. Proc Natl Acad Sci U S A 110:3507–3512

14. Takao K, Miyakawa T (2015) Genomic responses in mouse models greatly mimic human inflammatory diseases. Proc Natl Acad Sci U S A 112:1167–1172

15. Deitch EA (1998) Animal models of sepsis and shock: a review and lessons learned. Shock 9:1–11

16. Efron PA, Mohr AM, Moore FA, Moldawer LL (2015) The future of murine sepsis and trauma research models. J Leukoc Biol 98:945–952

17. Pratt D (1980) Alternatives to pain in experiments on animals. Argus Archives, New York

18. Tsukamoto T, Pape HC (2009) Animal models for trauma research: what are the options? Shock 31:3–10

19. Frink M, Andruszkow H, Zeckey C, Krettek C, Hildebrand F (2011) Experimental trauma models, an update. J Biomed Biotechnol 2011:797383

20. Noble RL, Collip JB (1942) A quantitative method for the production of experimental traumatic shock without haemorrhage in unanaesthetized animals. Q J Exp Physiol 31:187–199

21. Gill R, Ruan X, Menzel CL, Namkoong S, Loughran P, Hackam DJ, Billiar TR (2011) Systemic inflammation and liver injury following hemorrhagic shock and peripheral tissue trauma involve functional TLR9 signaling on bone marrow-derived cells and parenchymal cells. Shock 35:164–170

22. Kang SC, Matsutani T, Choudhry MA, Schwacha MG, Rue LW, Bland KI, Chaudry IH (2004) Are the immune responses different in middle-aged and young mice following bone fracture, tissue trauma and hemorrhage? Cytokine 26:223–230

23. Matsutani T, Kang SC, Miyashita M, Sasajima K, Choudhry MA, Bland KI, Chaudry IH (2007) Young and middle-age associated differences in cytokeratin expression after bone fracture, tissue trauma, and hemorrhage. Am J Surg 193:61–68

24. Matsutani T, Kang SC, Miyashita M, Sasajima K, Choudhry MA, Bland KI, Chaudry IH (2007) Liver cytokine production and ICAM-1 expression following bone fracture, tissue trauma, and hemorrhage in middle-aged mice. Am J Physiol Gastrointest Liver Physiol 292:G268–G274

25. Venet F, Chung CS, Huang X, Lomas-Neira J, Chen Y, Ayala A (2009) Lymphocytes in the development of lung inflammation: a role for regulatory CD4+ T cells in indirect pulmonary lung injury. J Immunol 183:3472–3480

26. Wichmann MW, Ayala A, Chaudry IH (1998) Severe depression of host immune functions following closed-bone fracture, soft-tissue trauma, and hemorrhagic shock. Crit Care Med 26:1372–1378

27. Wang P, Ba ZF, Burkhardt J, Chaudry IH (1993) Trauma-hemorrhage and resuscitation in the mouse—effects on cardiac-output and organ blood-flow. Am J Physiol 264:H1166–H1173

28. Hollenberg SM (2005) Mouse models of resuscitated shock. Shock 24(Suppl 1):58–63

29. Marshall JC, Deitch E, Moldawer LL, Opal S, Redl H, van der Poll T (2005) Preclinical models of shock and sepsis: what can they tell us? Shock 24(Suppl 1):1–6

30. Xiao W, Mindrinos MN, Seok J, Cuschieri J, Cuenca AG, Gao H, Hayden DL, Hennessy L, Moore EE, Minei JP et al (2011) A genomic storm in critically injured humans. J Exp Med 208:2581–2590

31. Gentile LF, Nacionales DC, Lopez MC, Vanzant E, Cuenca A, Cuenca AG, Ungaro R, Baslanti TO, McKinley BA, Bihorac A et al (2014) A better understanding of why murine models of trauma do not recapitulate the human syndrome. Crit Care Med 42:1406–1413

32. Yue F, Cheng Y, Breschi A, Vierstra J, Wu W, Ryba T, Sandstrom R, Ma Z, Davis C, Pope BD et al (2014) A comparative encyclopedia of DNA elements in the mouse genome. Nature 515:355–364

33. Cuschieri J, Johnson JL, Sperry J, West MA, Moore EE, Minei JP, Bankey PE, Nathens AB, Cuenca AG, Efron PA et al (2012) Benchmarking outcomes in the critically injured trauma patient and the effect of implementing standard operating procedures. Ann Surg 255:993–999

34. Chute CG, Ullman-Cullere M, Wood GM, Lin SM, He M, Pathak J (2013) Some experiences and opportunities for big data in translational research. Genet Med 15:802–809

35. Kannry JL, Williams MS (2013) Integration of genomics into the electronic health record: mapping terra incognita. Genet Med 15:757–760

36. Dyson A, Singer M (2009) Animal models of sepsis: why does preclinical efficacy fail to translate to the clinical setting? Crit Care Med 37:S30–S37

37. Keel M, Trentz O (2005) Pathophysiology of polytrauma. Injury 36:691–709

38. Gauthier C, Griffin G (2005) Using animals in research, testing and teaching. Rev Sci Tech 24:735–745

39. Gentile LF, Nacionales DC, Cuenca AG, Armbruster M, Ungaro RF, Abouhamze AS, Lopez C, Baker HV, Moore FA, Ang DN et al (2013) Identification and description of a novel murine model for polytrauma and shock. Crit Care Med 41:1075–1085

40. Mira JC, Szpila BE, Nacionales DC, Lopez MC, Gentile LF, Mathias BJ, Vanzant EL, Ungaro R, Holden D, Rosenthal MD et al (2016) Patterns of gene expression among murine models of hemorrhagic shock/trauma and sepsis. Physiol Genomics 48:135–144

41. Baker SP, O'Neill B, Haddon W Jr, Long WB (1974) The injury severity score: a method for describing patients with multiple injuries and evaluating emergency care. J Trauma 14:187–196

42. Ayala A, Chung CS, Lomas JL, Song GY, Doughty LA, Gregory SH, Cioffi WG, LeBlanc BW, Reichner J, Simms HH et al (2002) Shock-induced neutrophil mediated priming for acute lung injury in mice: divergent effects of TLR-4 and TLR-4/FasL deficiency. Am J Pathol 161:2283–2294

43. Lomas JL, Chung CS, Grutkoski PS, LeBlanc BW, Lavigne L, Reichner J, Gregory SH, Doughty LA, Cioffi WG, Ayala A (2003) Differential effects of macrophage inflammatory chemokine-2 and keratinocyte-derived chemokine on hemorrhage-induced neutrophil priming for lung inflammation: assessment by adoptive cells transfer in mice. Shock 19:358–365

44. Council NR (2011) Guide for thc care and use of laboratory animals, 8th edn. National Academies Press, Washington, DC

45. Cuenca AG, Delano MJ, Kelly-Scumpia KM, Moldawer LL, Efron PA (2010) Cecal ligation and puncture. In: Coligan JE et al (eds) Current protocols in immunology, Chapter 19, Unit 19 13. Wiley, New York

46. Stephan RN, Kuppr TS, Geha AS, Baue AE, Chaudry IH (1987) Hemorrhage without tissue trauma produces immunosuppression and enhances susceptibility to sepsis. Arch Surg 122:62–68

47. Monaghan SF, Thakkar RK, Heffernan DS, Huang X, Chung CS, Lomas-Neira J, Cioffi WG, Ayala A (2012) Mechanisms of indirect acute lung injury a novel role for the coinhibitory receptor, programmed death-1. Ann Surg 255:158–164

48. Bible LE, Pasupuleti LV, Gore AV, Sifri ZC, Kannan KB, Mohr AM (2015) Chronic restraint stress after injury and shock is associated with persistent anemia despite prolonged elevation in erythropoietin levels. J Trauma Acute Care Surg 79:91–97

49. Darwiche SS, Kobbe P, Pfeifer R, Kohut L, Pape HC, Billiar T (2011) Pseudofracture: an acute peripheral tissue trauma model. J Vis Exp (50):2074

50. Schwulst SJ, Trahanas DM, Saber R, Perlman H (2013) Traumatic brain injury-induced alterations in peripheral immunity. J Trauma Acute Care 75:780–788

51. Lpaktchi K, Mattar A, Niederbichler AD, Kim J, Hoesel LM, Hemmila MR, GL S, Remick DG, Wang SC, Arbabi S (2007) Attenuating burn wound inflammation improves pulmonary function and survival in a burn-pneumonia model. Crit Care Med 35:2139–2144

Chapter 2

Measurement of Intracranial Pressure in Freely Moving Rats

Michael R. Williamson, Roseleen F. John, and Frederick Colbourne

Abstract

Brain injury, such as from stroke and trauma, can be complicated by elevated intracranial pressure (ICP). Although raised ICP can be a significant determinant of morbidity and mortality, clinical studies often report widely varying ICP measurements depending on location of measurement and technique used. For the same reasons, reported ICP measurements also vary widely in animal models. The need for anesthesia or tethered connections with some methods of ICP measurement in animals may introduce additional confounds. Moreover, these methods are not well suited for prolonged, continuous measurement. Here, we describe an approach to continually measure ICP in awake, freely moving rats for several days. This technique uses a commercially available, wireless pressure sensor mounted on the head to measure ICP from the epidural space via a fluid-filled catheter. We have demonstrated that this approach reliably detects elevations in ICP that last for several days after ischemic and hemorrhagic strokes in rat.

Key words Stroke, Trauma, ICP, Intracranial pressure, Telemetry, Rodent, Animal model, Blood pressure sensor, Brain injury

1 Introduction

Elevated intracranial pressure (ICP) is a potentially deleterious and sometimes fatal response following brain injury such as stroke and trauma [1]. Since ICP is an important determinant and predictor of outcome [2, 3], its monitoring is sometimes used to guide clinical treatment [4]. Animal models are used to mechanistically understand ICP changes and to test potential therapies. As in patients, various ICP measurement protocols have been used in animals [5]; however, these methods often do not monitor ICP frequently or long enough to properly document the time course of ICP changes. Infrequent monitoring may miss brief but potentially significant ICP spikes [6]. Additionally, use of an anesthetic to immobilize animals potentially causes a sedative-induced ICP decline, thereby confounding measurements [7]. Further, certain anesthetics can impact cell death and recovery, making them less than ideal for use in stroke and trauma studies [8]. Finally, fluid-filled tethered lines used to measure ICP in nonanesthetized

Binu Tharakan (ed.), *Traumatic and Ischemic Injury: Methods and Protocols*, Methods in Molecular Biology, vol. 1717, https://doi.org/10.1007/978-1-4939-7526-6_2, © Springer Science+Business Media, LLC 2018

animals are prone to movement artefacts and restraint-induced stress, potentially confounding ICP measurements. Our method for wireless measurement of ICP in freely moving rats largely overcomes these obstacles.

Intraventricular [5, 9, 10], intraparenchymal [10], and epidural [5, 6, 11] locations are commonly used to measure ICP. However, catheter insertion to intraventricular and intraparenchymal locations causes tissue damage. Alternatively, measurement of ICP with epidural catheters has been reported as less invasive and more reliable [5], though some question its accuracy [12]. Importantly, epidural catheter placement avoids exacerbation of tissue damage that may occur as tissue swells around catheters inserted into the brain—swelling commonly occurs after stroke and trauma [13]. Unfortunately, there is little consistency or consensus regarding the location of ICP measurement [14].

Here, we describe a method for measurement of ICP in freely moving rats using a telemetric blood pressure transmitter connected to an epidural cannula. The probe is encased in a protective cylinder mounted onto the skull. This technique avoids potential confounds of using an anesthetic or physical restraint (tethering of the animal). Moreover, this method has been successfully used to monitor ICP for days after experimental ischemic stroke [11] and intracerebral hemorrhage in rats [6]. Lastly, the described technique itself causes less brain injury than many alternative methods, an important factor in stroke and trauma studies.

2   Materials

2.1   Head Assembly

1. Epidural cannula (*see* **Note 1** and Fig. 1a).

2. Hollow nylon screws (PlasticsOne, Roanoke, VA, model C212SGN; *see* **Note 2** and Fig. 1a).

3. Metal screws (Small Parts via Amazon Business, Seattle, WA, model B000FMUH4M, 1/8 in. length, #0-80 thread size; *see* **Note 3** and Fig. 1a).

4. PE 20 tubing (Instech Laboratories Inc., Plymouth Meeting, PA, 0.38 mm internal diameter, 1.09 mm external diameter).

5. Plastic cylinder (*see* **Note 4** and Fig. 1c).

2.2   Sterilization Equipment and Solutions

All surgical equipment and supplies must be sterilized prior to surgery.

1. Autoclave.

2. Ethanol (70%) (*see* **Note 5**).

3. Sterilant (*see* **Note 6**).

4. Topical antiseptic (*see* **Note 7**).

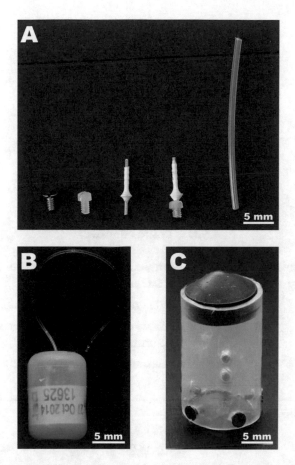

Fig. 1 Components of the head assembly are shown. (**a**) (From left to right) Metal screw, uncut hollow nylon screw, cut and filed 23 G needle (epidural cannula), cannula inserted into cut nylon screw, PE 20 tubing. (**b**) Telemetric pressure sensor. (**c**) Plastic cylinder with metal screws inserted around bottom circumference and rubber stopper to seal the top

2.3 Surgical Equipment and Supplies

1. Cotton swabs.
2. Dental cement.
3. Drill bit (0.7 mm diameter; *see* **Note 8**).
4. Electric shaver.
5. Eye lubricant.
6. Forceps.
7. Gauze.
8. General anesthetic suitable for rodent surgery (*see* **Note 9**).
9. Hemostats.
10. Local anesthetic (*see* **Note 10**).
11. Needles (23 G).

12. Rectal temperature probe connected to control unit and heating pad.

13. Rubber stopper.

14. Scalpel handle with size 11 blade.

15. Short screwdriver.

16. Stereotaxic drill.

17. Stereotaxic frame suitable for small animal surgery.

18. Sterile saline (*see* **Note 11**).

19. Surgical drapes.

20. Syringes (1 mL).

21. Tissue adhesive.

2.4 Telemetry Equipment

1. Ambient Pressure Reference (Data Sciences Int., St. Paul, MN, Model APR-1).

2. Data collection software (Data Sciences Int., Dataquest A.R.T.).

3. Pressure probe (Data Sciences Int., Model PA-C10; *see* **Note 12** and Fig. 1b).

4. Telemetry receiver (Data Sciences Int., Model RPC-1).

3　Methods

Surgical procedures should be performed aseptically.

3.1 Surgical Procedures

1. Turn on ICP probe (*see* **Note 13**).

2. Anesthetize animal (*see* **Note 9**).

3. Shave dorsal aspect of head. Clear area of loose hairs with ethanol and gauze.

4. Inject local anesthetic subcutaneously in center of shaved area (incision site; *see* **Note 10**).

5. Transfer animal to stereotaxic frame (*see* **Note 14**).

6. Insert rectal temperature probe and maintain normothermia throughout procedure.

7. Apply eye lubricant.

8. Clean shaved area with 70% ethanol then topical antiseptic (*see* **Note 7**).

9. Drape animal.

10. Make a midline scalp incision (approximately 2.5 cm).

11. Scrape away membrane from skull using cotton swabs.

Fig. 2 Screws inserted into skull with metal cannula inserted into epidural space through central nylon screw

12. Drill a central hole for catheter insertion and surrounding holes for metal screws to stabilize the head assembly (*see* **Note 15** and Fig. 2).

13. Gently widen holes using drill bit. Insert metal screws into surrounding holes to inner skull surface.

14. Fill hollow nylon screw with sterile saline and insert into central burr hole (*see* **Note 16**). Apply tissue adhesive to further secure screw. Take care to avoid blocking hollow center with adhesive.

15. Attach a 3 cm piece of PE 20 tubing to cannula (*see* Fig. 1a). Fill with sterile saline. Ensure that no bubbles are present within tubing and catheter.

16. Insert the catheter with attached tubing into hollow screw. Secure with adhesive.

17. Insert lead of pressure probe into tubing at end opposite catheter (*see* **Note 17**).

18. Guide pressure probe through the plastic cylinder, and center the plastic cylinder over the hollow screw.

19. Insert metal screws into cylinder holes (*see* **Note 4** and Fig. 2).

20. Secure cylinder with dental cement (*see* **Note 18**).

21. Once the dental cement hardens, place ICP probe within cylinder and seal with rubber stopper. Take care that tubing does not kink.

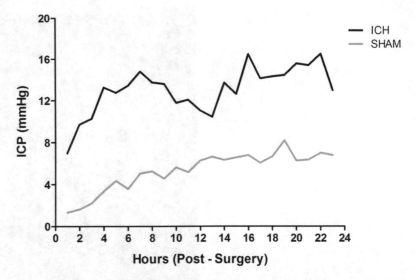

Fig. 3 Sample traces of intracranial pressure (ICP, mmHg) are shown for the first 24 h following either an intracerebral hemorrhage (ICH) or a sham stroke in two rats. Infusing bacterial collagenase into the striatum produced the ICH, whereas only saline was infused in the sham animal (no stroke). These sample data from two adult male rats were collected as part of study examining ICP responses following untreated and treated ICH. The biosciences Animal Care and Use Committee at the University of Alberta approved this work and procedures

22. Discontinue anesthetic and hydrate animal subcutaneously with 5 mL of sterile saline.
23. Place animal in home cage (*see* **Note 19**) on top of telemetry receiver to begin recordings (*see* **Note 20** and Fig. 3).
24. Provide appropriate postoperative care. Ensure that head assembly does not become dislodged (*see* **Note 21**).

4   Notes

1. Any cannula can be used provided it fits the internal diameter of the PE 20 tubing (0.38 mm, though this can be slightly widened) and the nylon screw (0.50 mm). A 23 G needle shaft can be cut and filed to create an appropriately sized cannula. Care should be taken to ensure that the needle shaft does not exceed the length of the nylon screw (*see* Fig. 1a). A 3 cm length of sterile saline-filled (*see* **Note 11**) PE20 tubing is connects the pressure probe catheter with the epidural cannula. The pressure probe catheter should not be directly attached to the cannula to avoid damage to probe. Likewise, care must be taken to avoid exerting excessive positive or negative pressure on the telemetry probe, which may result in probe failure. Manufacturer guidelines should be followed.

2. The shaft of the nylon screw should be cut to fit the thickness of the rat's skull (approximately cut in half if PlasticsOne model C212SGN, *see* Fig. 1a). Cannula should be threaded through the hole of the nylon screw prior to surgery to ensure it fits well.

3. Screw length should not exceed 1/8 in. Small Parts model B000FMUH4M screws work well for rat skull. Care should be taken that screws are not inserted past the inner skull surface to avoid injuring the brain.

4. To contain and protect the telemetry probe from damage, a 2.5 cm long cylinder can be cut from a 5 mL syringe barrel (12 mm internal diameter, size of cylinder can be changed depending on size of probe). The top of the cylinder must be sealed with a rubber stopper. Screws are inserted around the lower diameter of the cylinder to help secure cylinder to dental cement (*see* Fig. 1c).

5. Screws, catheters, and PE 20 tubing must be sterilized prior to use. Tubing cannot be autoclaved. Cidex OPA and ethylene oxide are alternatives to ethanol that provide better sterilizing action. All instruments must be kept on sterile nonporous drapes during use.

6. We submerge the telemetry probe in 2% glutaraldehyde for a minimum of 12 h, then thoroughly rinse several times with sterile saline prior to use.

7. We apply Betadine to shaved head three times after cleaning with 70% ethanol.

8. A 0.7 mm diameter drill bit is used to slightly widen burr holes before inserting screws. This ensures a tight seal and easier insertion.

9. We use isoflurane (4% induction, 2–3% maintenance) in 60% N_2O and 40% O_2.

10. Subcutaneously inject local anesthetic (e.g., 0.1 mL of Marcaine) prior to incision. Local anesthetics can also be combined with systemic analgesics to control pain in rats, but care must be taken to avoid such agents that might affect ICP. Alternatives to Marcaine include lidocaine and ropivaciane.

11. Alternatively, sterile artificial cerebrospinal fluid should be used.

12. Alternative models are available (HD-S10 and PA-C40, Data Sciences Int.), but are larger and heavier. Model PA-C10 weighs 1.4 g. The total weight of the head assembly and probe is approximately 2.7 g. We implant this probe on rats weighing at least 350 g. Use of larger probes is not recommended.

13. Pressure readings may take several hours to normalize after surgery owing to opening the skull during the surgical procedure. Pressure probe should be turned on several hours prior to surgery to stabilize its readings. Baseline recordings should be taken within 24 h prior to surgery to account for pressure drift, which may confound ICP readings. However, drift is typically very small (~0.25 mmHg per month on average).

14. Other invasive procedures (e.g., intraparenchymal injections) should be done prior to implantation of pressure probe. This technique requires the skull to be sealed and largely precludes other procedures.

15. At least three stabilizing screws should be used (*see* Fig. 2). Head of metal screws should border the inner lining of the plastic cylinder, which should be centered over the catheter. We drill the centre burr hole 3.5 mm to the right and 0.07 mm posterior of Bregma. Other locations can be used provided the plastic cylinder can be securely mounted on the skull.

16. Threading the center hole with a metal screw inserted to the inner surface of the skull, and then removed, makes insertion of the nylon screw much easier.

17. The pressure probe catheter is very sensitive. Air bubbles, blood components, or foreign materials in the pressure-sensing region may cause inaccurate pressure readings. Attachment of the probe catheter directly to epidural catheter would damage the probe. Care must be taken when inserting, removing, and cleaning the catheter tip to avoid damage. The catheter tip must be level with the body of the transmitter when taking pressure offsets. Inaccurate pressure recordings after implant may be due to damage sustained to the transducer during insertion. System integrity can be confirmed by performing a brief abdominal compression (~1 s, *see* ref. 11). This should cause a transient ICP spike.

18. Dental cement should be applied both to interior of cylinder to secure it to metal screws and exterior to seal incision wound and further secure cylinder. Additional local anesthetic can be applied prior to this step.

19. Animals should be individually housed and objects should be removed to prevent dislodging of head assembly. Cages can be fitted with high top cage lids to create additional height and allow upright standing without interfering with head apparatus.

20. Using this method, the average ICP in uninjured rats is approximately 5 to 6 mmHg (*see* refs. 6, 11). It is important to note that ICP readings vary considerably across different ICP monitoring methods and/or locations (*see* refs. 5, 9, 10). We

strongly recommend using appropriate controls in each experiment (e.g., stroke vs. sham-operated rats).

21. It is possible for the head assembly to dislodge, potentially requiring the animal to be euthanized and risking damage to the telemetry probe. Probes are very delicate and can be easily damaged by the animal if the head assembly becomes dislodged. Care must be taken to adequately monitor rats for postoperative pain, dehydration, and malnourishment and to ensure the head assembly is well secured. Soft food, such as peanut butter, may be given to encourage eating and improve hydration as determined by veterinarian consultation. Oral hydration gels or intraperitoneal saline can also be given.

References

1. Steiner T, Weber R, Krieger D (2001) Increased intracerebral pressure following stroke. Curr Treat Options Neurol 3:441–450

2. Steiner LA, Andrews PJD (2006) Monitoring the injured brain. ICP and CBF. Br J Anaesth 97:26–38

3. Sykora M, Steinmacher S, Steiner T et al (2014) Association of intracranial pressure with outcome in comatose patients with intracerebral hemorrhage. J Neurol Sci 342:141–145

4. Kirkman MA, Smith M (2014) Intracranial pressure monitoring, cerebral perfusion pressure estimation, and ICP/CPP-guided therapy: a standard of care or optional extra after brain injury? Br J Anaesth 112:35–46

5. Uldall M, Juhler M, Skjolding AD et al (2014) A novel method for long-term monitoring of intracranial pressure in rats. J Neurosci Methods 227:1–9

6. Hiploylee C, Colbourne F (2014) Intracranial pressure measured in freely moving rats for days after intracerebral hemorrhage. Exp Neurol 255:49–55

7. Flower O, Hellings S (2012) Sedation in traumatic brain injury. Emerg Med Int 2012:1–11

8. Schifilliti D, Grasso G, Conti A, Fodale V (2010) Anaesthetic-related neuroprotection: intravenous or inhalational agents? CNS Drugs 24:893–907

9. Chowdhury UR, Holman BH, Fautsch MP (2013) A novel rat model to study the role of intracranial pressure modulation on optic neuropathies. PLoS One 8:1–8

10. Zwienenberg M, Gong QZ, Lee LL et al (1999) ICP monitoring in the rat: comparison of monitoring in the ventricle, brain parenchyma, and cisterna magna. J Neurotrauma 16:1095–1102

11. Silasi G, MacLellan CL, Colbourne F (2009) Use of telemetry blood pressure transmitters to measure intracranial pressure (ICP) in freely moving rats. Curr Neurovasc Res 6:62–69

12. Poca MA, Sahuquillo J, Topczewski T et al (2007) Is intracranial pressure monitoring in the epidural space reliable? Fact and fiction. J Neurosurg 106:548–556

13. MacLellan CL, Silasi G, Poon CC et al (2008) Intracerebral hemorrhage models in rat: comparing collagenase to blood infusion. J Cereb Blood Flow Metab 28:516–525

14. Olson DM, Batjer HH, Abdulkadir K, Hall CE (2013) Measuring and monitoring ICP in neurocritical care, results from a national practice survey. Neurocrit Care 20:15–20

Chapter 3

A Lateral Fluid Percussion Injury Model for Studying Traumatic Brain Injury in Rats

Paige S. Katz and Patricia E. Molina

Abstract

Traumatic brain injury (TBI) diagnoses have increased in frequency during the past decade, becoming a silent epidemic. The pathophysiology of TBI involves pathophysiological processes affecting the brain, induced by traumatic biomechanical forces resulting in temporary impairment of neurological function. Preclinical models have been generated to recapitulate the mechanical, neuroinflammatory, and behavioral outcomes observed in the clinical setting. The lateral fluid percussion (LFP) model is the most extensively used and well-characterized model of nonpenetrating and nonischemic TBI. The model is reproducible and can be adjusted to produce a mild to moderate and severe injury, as reflected by mortality and return of reflexes, by adjusting the amount of force applied. The histopathological changes achieved with this model reproduce that seen in human TBI including focal contusion in the cortex, with accompanying intraparenchymal punctate hemorrhage, followed by inflammation and neuronal degeneration. This chapter describes the LFP model, which produces a mixed model of focal and diffuse brain injury that progresses over time affecting predominantly the cortical parenchyma.

Key words Traumatic brain injury, Fluid percussion injury, Lateral, Craniotomy

1 Introduction

Traumatic brain injury (TBI) is a national healthcare problem with an increasing frequency, affecting an estimated 1.7 million Americans per year [1, 2]. Mild TBI and concussions are the most common form of TBI, accounting for nearly 88% of all TBI related incidents, and TBI is a leading cause of morbidity and mortality in young adults in the USA [1, 3]. Currently, TBI is a poorly recognized and understood injury, as the majority of patients appear to recover within days to weeks. However, TBI is a complex process involving the primary injury or mechanical insult, followed by the secondary injury which can manifest over a period of hours to days and even years [4]. The processes and mechanisms underlying the secondary injury are the focus of intense investigation targeted toward identifying the molecular mechanisms involved and

Binu Tharakan (ed.), *Traumatic and Ischemic Injury: Methods and Protocols*, Methods in Molecular Biology, vol. 1717,
https://doi.org/10.1007/978-1-4939-7526-6_3, © Springer Science+Business Media, LLC 2018

providing potential therapeutic treatments to reduce long term damage following the initial injury [5–7].

Several experimental models have been developed to reproduce and study the pathological sequelae following mild TBI or concussions in humans [8–11]. Initially, the midline fluid percussion model was developed to study brain injury, and has since been modified to the lateral fluid percussion (LFP) for use in rodents [8, 9]. The LFP model of TBI is an extensively used, well-accepted, and reliable method for reproducing several pathological responses as seen clinically [12–14].

2 Materials

2.1 Surgery

1. Stereotactic alignment frame (Model 900, KOPF Instruments) for rats, with non-rupture ear bars, and gas anesthesia nose holder (KOPF Instruments).

2. Isoflurane vaporizer/anesthesia machine (Parkland Scientific) and induction chamber (World Precision Instruments).

3. F/Air scavenger carton (Braintree Scientific, Inc.).

4. A warming pad placed under the rat and set at 37 °C to maintain constant body temperature during surgery.

5. Maxi Cide/Glutaraldehyde (Henry Schein).

6. Sterilized/autoclaved water.

7. Clippers (Oster).

8. Betadine surgical scrub (povidone–iodine, 7.5%) and isopropyl alcohol.

9. Surgical gooseneck light.

10. Artificial tears (Henry Schein).

11. Sterile gloves, face mask, and hair bonnet.

12. Sterilized instruments pictured (*see* Fig. 1 Scalpel blade (#22) and holder (Becton Dickinson), Hemostats (Halstead Mosquito Forceps 5″ Curved, 1.3 mm tips, Roboz, Gaithersburg, MD), Michele Trephine, 5 mm outer diameter (6.25 3/16″ OD, Roboz), Fine forceps #7 (Dumont #7 Vessel Dilating Forceps Inox Tip Size 0.20 × 0.16 mm, Roboz).

13. Sterilized instruments not pictured: Bone scraper (Spratt Curette 6.5″ Oval Size 2/0, Roboz), stainless steel mounting screws (Size 0–80 × 3/32, Plastics One, Roanoke, VA).

14. Sterilized cotton tipped applicators and 2 × 2 gauze.

15. 20G 1″ needles (Becton Dickinson) with the needle cut off to be used as the cannula over the craniotomy (*see* Fig. 2).

16. Sterile saline.

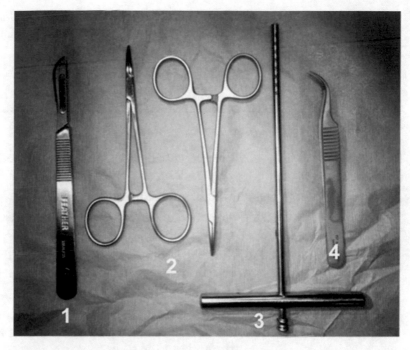

Fig. 1 Picture of instruments used for craniotomy surgery. (1) Scalpel holder and scalpel blade, (2) Hemostats, (3) Michele trephine, (4) Fine forceps

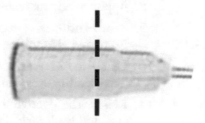

Using a razor blade, cut approximately 3mm behind needle connection.

Fig. 2 Example of cannula, made from a 20G 1″ needle

17. Three way stop cocks for caps.

18. Super glue gel, jet denture repair powder and liquid (Lang Dental).

2.2 Lateral Fluid Percussion

1. High pressure tubing with male Luer-lock attachment (Baxter, Extension Set).

2. Fluid percussion instrument (FPI) with pressure transducer amplifier (Model # 01-B, Custom Design and Fabrication, Virginia Commonwealth University).

3. Computer (PC or Mac) and Lab Chart for data acquisition (ADInstruments, Colorado Springs, CO).

4. Sterile water.

5. 25 mL syringe.

3 Methods

*3.1 Surgery/
Craniotomy*

1. Prepare sterile operating field with sterile drape and open surgery pack on to sterile field.

2. Weigh and record rat's weight.

3. Anesthetize rat with 3–4% isoflurane for 4–5 min in the induction chamber.

4. Shave the animals head from behind the ears to the middle of the snout and remove any loose hairs.

5. Place back in induction chamber for 2–3 min.

6. Move rat to stereotactic frame and redirect isoflurane to nose holder on stereotactic frame. Maintain anesthesia at 2–3% throughout the surgery.

7. Stabilize the head in a stereotaxic frame using the incisor bar and a pair of nonrupture ear bars.

8. Apply an adequate amount of eye lubricant to each eye.

9. Scrub and clean surgical area first with Betadine followed by Isopropyl alcohol; repeat three times.

10. Using a #22 scalpel blade, make an incision (~3 cm) from between the eyes and over the cranial crown.

11. Reflect the overlying skin and muscle to expose the cranium and pull back each side with curved hemostats.

12. Using sterile cotton tipped applicators, swab the skull to remove any membrane.

13. Using the bone scrapper, gently scrape the skull in both directions to prepare the surface for super glue and dental acrylic (*see* **Note 1**).

14. Place a 2 × 2 piece of gauze over skull and leave for a few minutes to ensure that the skull is completely dry.

15. Hand drill one hole posterior to Lambda, to place an anchor screw that will help secure the dental acrylic.

16. Carefully set the screw, making certain not to contact the dura (three full turns is sufficient).

17. Using the stereotaxic frame, locate bregma (*see* Fig. 2) and position the left arm of the stereotaxic frame in the center of bregma and the lateral suture.

18. Using the dials on the stereotactic frame, move the left arm 2 mm posterior (back) to the bregma, and 3 mm lateral (left) to midline.

19. Mark the skull with a permanent marker to indicate the site of the drill hole.

20. Using the Michele Trephine gently drill a circular track with the mark from the **step 18** as its center (*see* **Note 2**).

21. Drill until the skull flap depresses with light force by forceps.

22. Remove the circular skull flap. Be sure the dura is intact before continuing (*see* **Note 3**).

23. Using sterile gauze, wipe the skull of any bone fragments and ensure the skull is completely dry before continuing.

24. Introduce the head cannula (make sure that it is completely dry) to the craniotomy hole until it is directly above the dura.

25. The cannula should fit the craniotomy hole snugly and there should be nothing obscuring visualization of the dura through the center of the cannula.

26. Once you have aligned the cannula directly over the drill hole, seal the perimeter with super glue.

27. Allow the glue to dry completely for approximately 5 min (*see* **Note 4**).

28. Fill the cannula with sterile saline to ensure the dura does not dry out.

29. Mix equal parts of dental acrylic and jet liquid in a small tray. The amount needed should cover the exposed skull and surround the cannula (*see* **Note 5**).

30. Using a 1 mL syringe, draw up the acrylic mixture and apply one layer of dental acrylic glue (*see* **Note 6**). Allow the first layer to dry and then apply a second layer.

31. Once the dental acrylic is completely dry, apply one final layer of Super Glue® around the cannula sealing the cannula to the dental acrylic for additional reinforcement.

32. Once the glue is dry, cap the cannula to keep the dura moist during recovery from surgery.

33. Remove the animal from the stereotaxic head frame and place on its side, in a clean cage (which should sit on a heating pad) for observations until the animal regains consciousness (when using isoflurane the animal typical wakes up within 5–10 min after being place in the recovery cage).

3.2 Induction of TBI Using a Fluid Percussion Instrument

1. To set up the FPI fill the chamber with sterile water making sure that no bubbles are present (*see* **Note 7**).

2. Attach a 25 mL syringe filled with sterile water to the center port and use this to remove any bubbles and/or fill the chamber after use.

3. Prior to use, confirm that the FPI and the high-pressure tubing connected to it are filled with sterile water and free of air bubbles.

4. Turn on the pressure transducer amplifier that is connected to the FPI.

5. Calibrate the instrument using the 100 PSI calibration switch. A two point calibration can be completed by setting the pressure in pounds per square inch (PSI), with the pressure tubing open, to the zero level set your value to zero in you data acquisition software; to set the 100 level, switch on the 100 PSI calibration switch and set that value to 100 in your data acquisition software.

6. To complete the setup, turn the data acquisition on and with the Luer-lok end of the tubing closed, release the pendulum to deliver test pulses. The device should be primed by delivering approximately five test pulses. Confirm that pendulum gives a smooth signal on the oscilloscope and amplifier (*see* Fig. 3). A noisy signal indicates air in the system that must be removed prior to delivering the injury pulse (*see* **Note 8**).

7. If necessary, adjust the angle of pendulum to increase or decrease the intensity of the pulse. The angle of the pendulum's starting position for a mild to moderate TBI is approximately 18–20° (*see* **Note 9**).

8. Placed the rat in the 4–5% isoflurane anesthesia chamber (pre-charged) until a surgical plane of anesthesia is reached. The rat is placed on the stereotaxic frame next to the FPI and the hub cleaned of any clotted blood and then filled with sterile saline. The tubing of the FPI with a male Luer-lok is attached to the female Luer-lok fitting of the cannula hub.

9. Turn the isoflurane off, once a normal breathing pattern resumes but prior to the animal regaining complete consciousness (~1–2 min), release the pendulum of the FPI to cause a single pulse of injury (Fig. 4). It is important to not induce the injury while the animal is too anesthetized as it could cause severe respiratory depression and death. The exact pressure of the pulse should be recorded. Uninjured, sham animals undergo all of the same procedures with the exception of the actual fluid pulse to induce injury.

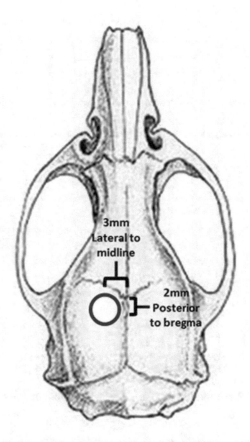

Fig. 3 Location of craniotomy on the left side of the rat skull

10. Start a timer at the time of injury so that apnea and righting reflex can be recorded (see below). Immediately after the injury each animal should be placed on the right side and allowed to recover.

    (a) For apnea, observe the time it takes the animal to take its first breath following injury. For sham animals, this will be zero.

    (b) For righting reflex, record the time it takes the animal to right itself on all four paws after being placed on its side.

11. Respiratory rate should also be taken and recorded following the injury.

12. Once the animal has righted to all fours paws, place it back in its cage and monitor the animals for at least 1 h post-TBI.

13. Follow-up behavioral assessments may be used to confirm injury severity and treatment effectiveness.

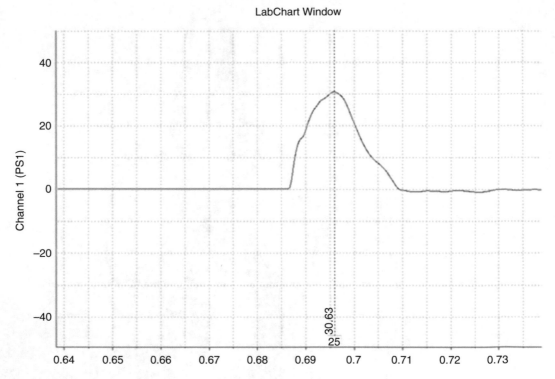

Fig. 4 Example wave form of a mild TBI (~30 PSI), obtained from LabChart

4   Notes

1. Scraping the skull can be done once, then wiped with gauze and repeated. Using this technique will improve the adherence of the super glue as well as the acrylic.

2. This step requires even pressure to create a groove. Once the groove is made, special care should be taken when applying downward pressure as the rat's skull is thin and too much pressure will damage the dura.

3. The inner dura has a distinct pink appearance and any damage to the dura will cause bruising or bleeding.

4. It is recommended that several thin layers of glue be applied, letting each layer dry completely before applying the next. Make sure the cannula and skull surface are completely dry before applying the glue, as fluid and tissue can cause the glue to not adhere properly.

5. Make sure that the acrylic is not too high up the cannula as this could impede securing of the cap in **step 30**.

6. Make sure to completely cover the anchor screw and the bottom half of the cannula. This will ensure that the cannula stays

in place during the injury. ALLOW THE ACRYLIC TO COMPLETELY DRY.

7. Directions on the setup and filling of the FPI are included with the instrument and should be followed prior to TBI.

8. The amplifier provided with the FPI converts the signal from the pressure transducer bridge to an output of 10 mV for every pound per square inch (PSI) relative to atmospheric pressure. To record the pressure for each event, the amplifier will need to be connected to either a computer with data acquisition software or an oscilloscope.

9. The transducer amplifier supplied with the FPI is calibrated such that 10 mV = 1.0 PSI. One atmosphere (ATM) = 14.7 PSI. Injury pressures delivered are typically in the range of 0.9–2.1 atmospheres to produce a range of righting reflex times and an increasing mortality associated with pulmonary edema. With this model, mild injury results in an average righting reflex time of 2–4 min and a 0–5% mortality rate. Moderate injury results in an average righting reflex time of 6–10 min and a 10–20% mortality rate.

Acknowledgments

We would like to acknowledge Jesse Sulzer, MD, PhD; Sophie Teng, MD, PhD; Renata Impastato; Emily Rogers; and Kylie Mills who have contributed to the refinement of this model.

References

1. Faul M, Xu L, Wald M, Coronado V (2010) Traumatic brain injury in the United States, Emergency Department visits, hospitalizations and deaths 2002–2006. Center for Disease Control and Prevention, National Center for Injury Prevention and Control, Atlanta, GA

2. Feinstein A, Rapoport M (2000) Mild traumatic brain injury: the silent epidemic. Can J Public Health 91:325–326. 32

3. Pellman EJ, Powell JW, Viano DC, Casson IR, Tucker AM, Feuer H, Lovell M, Waeckerle JF, Robertson DW (2004) Concussion in professional football, epidemiological features of game injuries and review of the literature—part 3. Neurosurgery 54:81–94. discussion -6

4. Werner C, Engelhard K (2007) Pathophysiology of traumatic brain injury. Br J Anaesth 99:4–9

5. Konrad C, Geburek AJ, Rist F, Blumenroth H, Fischer B, Husstedt I, Arolt V, Schiffbauer H, Lohmann H (2011) Long-term cognitive and emotional consequences of mild traumatic brain injury. Psychol Med 41(6):1197–1211

6. Gentleman SM, Leclercq PD, Moyes L, Graham DI, Smith C, Griffin WS, Nicoll JA (2004) Long-term intracerebral inflammatory response after traumatic brain injury. Forensic Sci Int 146:97–104

7. Riggio S (2011) Traumatic brain injury and its neurobehavioral sequelae. Neurol Clin 29:35–47

8. Dixon CE, Lyeth BG, Povlishock JT, Findling RL, Hamm RJ, Marmarou A, Young HF, Hayes RL (1987) A fluid percussion model of experimental brain injury in the rat. J Neurosurg 67:110–119

9. McIntosh TK, Vink R, Noble L, Yamakami I, Fernyak S, Soares H, Faden AL (1989) Traumatic brain injury in the rat, characterization of a lateral fluid-percussion model. Neuroscience 28:233–244

10. Dixon CE, Clifton GL, Lighthall JW, Yaghmai AA, Hayes RL (1991) A controlled cortical impact model of traumatic brain injury in the rat. J Neurosci Methods 39:253–262

11. Marmarou A, Foda MA, van den Brink W, Campbell J, Kita H, Demetriadou K (1994) A new model of diffuse brain injury in rats. Part I, pathophysiology and biomechanics. J Neurosurg 80:291–300

12. Thompson HJ, Lifshitz J, Marklund N, Grady MS, Graham DI, Hovda DA, McIntosh TK (2005) Lateral fluid percussion brain injury, a 15-year review and evaluation. J Neurotrauma 22:42–75

13. Povlishock JT (1993) Pathobiology of traumatically induced axonal injury in animals and man. Ann Emerg Med 22(6):980

14. Ling GSF, Lee EY, Kalehua AN (2001) Traumatic brain injury in the rat using the fluid-percussion model. Current protocols in neuroscience. Wiley, New York

Chapter 4

A Mouse Controlled Cortical Impact Model of Traumatic Brain Injury for Studying Blood–Brain Barrier Dysfunctions

Himakarnika Alluri, Chinchusha Anasooya Shaji, Matthew L. Davis, and Binu Tharakan

Abstract

Traumatic brain injury (TBI) is one of the leading causes of death and disability worldwide. It is a silently growing epidemic with multifaceted pathogenesis, and current standards of treatments aim to target only the symptoms of the primary injury, while there is a tremendous need to explore interventions that can halt the progression of the secondary injuries. The use of a reliable animal model to study and understand the various aspects the pathobiology of TBI is extremely important in therapeutic drug development against TBI-associated complications. The controlled cortical impact (CCI) model of TBI described here, uses a mechanical impactor to inflict a mechanical injury into the mouse brain. This method is a reliable and reproducible approach to inflict mild, moderate or severe injuries to the animal for studying TBI-associated blood–brain barrier (BBB) dysfunctions, neuronal injuries, brain edema, neurobehavioral changes, etc. The present method describes how the CCI model could be utilized for determining the BBB dysfunction and hyperpermeability associated with TBI. Blood–brain barrier disruption is a hallmark feature of the secondary injury that occur following TBI, frequently associated with leakage of fluid and proteins into the extravascular space leading to vasogenic edema and elevation of intracranial pressure. The method described here focuses on the development of a CCI-based mouse model of TBI followed by the evaluation of BBB integrity and permeability by intravital microscopy as well as Evans Blue extravasation assay.

Key words Blood–brain barrier, Controlled cortical impact, Traumatic brain injury, Intravital microscopy, Evans Blue, Edema, Intracranial pressure, Central nervous system, Hyperpermeability, Endothelial permeability

1 Introduction

Traumatic brain injury (TBI) is a complex injury occurs when an external mechanical force causes disruption in the normal function of the brain. Although there are multiple factors that contribute to the high mortality and morbidity observed in TBI, the development of cerebral edema, remains the most significant predictor of outcome. One of the major routes for the leakage of fluid and proteins from the intravascular space to the extravascular area "vascular hyperpermeability" leading to this edema is the blood–brain

Binu Tharakan (ed.), *Traumatic and Ischemic Injury: Methods and Protocols*, Methods in Molecular Biology, vol. 1717,
https://doi.org/10.1007/978-1-4939-7526-6_4, © Springer Science+Business Media, LLC 2018

barrier. Endothelial hyperpermeability is a significant problem in vascular conditions associated with traumatic and ischemic injuries, sepsis, atherosclerosis, acquired respiratory syndrome, diabetic retinopathy, thrombosis, inflammatory bowel disease, and cancer [1–4]. Of them, the most common traumatic injuries is traumatic brain injury (TBI). According to most recent CDC estimates (2010), there are 2.5 million new cases of TBI annually, with 50,000 deaths, 28,000 hospitalizations, and 2.2 million emergency department visits (http://www.cdc.gov/traumaticbraininjury). The large economic cost burden (direct and indirect costs) for healthcare costs, lost wages/productivity for patients suffering with TBI is estimated to be $76.5 billion according to 2012 CDC estimates [5, 6]. Disabilities that occur as a consequence of TBI include: emotional and mental disabilities; speech and language problems; loss of hearing, vision, taste, or smell; seizures; paralysis; and coma in some cases, and the chief causes of these injuries include violence, slips and falls, motor vehicle accidents, firearms, construction, sports, and other transports [7, 8].

TBI can be described as damage to brain due to an external mechanical force (rapid acceleration/deceleration, blast waves, crush or penetration by a projectile [9, 10]). TBI is divided into two main types of injuries: primary injuries due to a direct and immediate mechanical disruption of brain tissue by shear stress; and secondary injuries, a matrix of delayed events affecting neuronal, glial, and vascular structures leading to brain edema, BBB breakdown, and cerebral ischemia [11]. Primary injury that occurs after TBI cannot be treated but it can only be prevented. However, secondary injury that occurs following TBI can range from minutes to days and understanding the secondary injuries is critical for providing adequate treatment [12, 13]. TBI is clinically evaluated in patients with the help of four parameters: (1) Duration of loss of consciousness, (2) GCS scale (mild = GCS > =13; moderate = GCS 9–12; severe = GCS 3–9), (3) length of posttraumatic amnesia, and (4) ranchos Los amigos scale (1–10) [14, 15]. However, clinical symptoms of TBI may not be relevant to available animal models of TBI. The best way to understand the pathobiology of TBI and to identify potential therapeutic targets, is to generate relevant equivalents in behavior based on the assumed brain regions that are involved, and to find the underlying mechanisms even though precise mechanisms vary between animals and humans [16]. Most of the currently used animal models involve neuronal death and white matter damage that occur as part of both primary and secondary injuries in brain [12]. However, there are also several limitations that exist with the current animal models, like the physiological differences observed (due to differences in brain structure and function in the groups), gender differences (the possibility that females suffer less morbidities and complications), physiological variables like blood pressure, blood/brain ionic concentrations, partial pressures of carbon dioxide and oxygen, brain temperature, and accuracy of injury severity measurement, as even a small variation in the position of craniotomy can lead to variable results in cognitive abilities [16]. The CCI approach described here addresses several of these aspects including BBB dysfunctions.

The blood–brain barrier acts as a dynamic interface between the systemic blood circulation and the central nervous system. It acts as a barrier by regulating the exchange of substances and thus maintaining the normal neuronal activity in CNS. The BBB consists of brain endothelial cells along with various brain derived cells including astrocytes, pericytes, neurons, and extracellular matrix (ECM) and collectively comprises the neurovascular unit (NVU). The key feature of the BBB includes the well-defined tight junctions, comprising tight junction proteins (TJPs). TJPs play an important role in regulating the paracellular microvascular permeability. Permeability is shown to be largely influenced by various parameters like size, shape, charge, type of the extravasating molecules, nature of the vascular beds, etc. [17]. Tight junctions predominantly comprise of integral transmembrane proteins (claudins, occludin, junctional adhesion molecules, etc.) and accessory cytoplasmic membrane proteins (e.g., Zonula occludens-1; ZO-1). Brain microvascular hyperpermeability that occurs as a consequence of breach in the BBB following TBI is shown to be the chief cause of unfavorable outcomes like edema and elevated intracranial pressure (ICP) [18, 19]. Vasogenic edema is shown to be the most significant form of edema that occurs via the BBB and it results from the excessive leakage of plasma fluids and proteins resulting in microvascular hyperpermeability. In the meantime, an increase in the ICP that occurs as a result of this edema decreases the cerebral perfusion pressure (CPP), and thus, alters the cerebral blood flow affecting tissue oxygenation etc. The methods described in this chapter utilizes fluorescein-isothiocyanate (FITC) tagged dextran of known molecular weight (10 kDa) and Evans Blue dye, a dye when injected into blood, binds to serum albumin. The crossing of Evans Blue-bound albumin across the BBB is considered strong indications of BBB dysfunctions and microvascular hyperpermeability following TBI. The CCI technique described here is originally developed by James Lighthall and his colleagues in the 1980s. Since its development, it has become one of the most commonly used models of brain injury in multiple species of animals in laboratory settings. Furthermore, this method allows for the quantitative control over injury force and velocity as well as tissue deformation and the flexibility of use as it can be controlled easily in a research lab setting to produce a wide range of injury magnitudes and also produce gradable functional impairment, tissue damage, or both [16]. The main advantages of this method are high reproducibility, adjustable position and angle features, relatively low mortality rate of animals, and controlled deformation parameters like time, velocity, depth of impact with no rebound injury compared to other existing models. CCI mimics a wide range of focal and diffuse axonal injury which is most closest representation of clinical brain injury while the limitation being the need for craniotomy surgery, cost of the instrument, and complexity [12, 16].

2 Materials

2.1 Reagents

- 70% Alcohol.
- Artificial cerebrospinal fluid.
- Trichloroacetic acid.
- 0.9% Saline.
- Phosphate buffered saline (PBS).
- Betadine solution.

2.2 Equipment

- Gaymar pump.
- Heating pad.
- Intravital microscope.
- Control cortical impact.
- Cooling centrifuge.
- Homogenizer.
- Electric shaver.
- Micro drill with diamond tip.
- Burrs for micro drill.
- Timer.

2.3 Supplies

- Syringes.
- 25G, 27G/30G needles.
- Blunt and sharp end scissors.
- Blunt and sharp end forceps.
- Sterile cotton tipped applicators.
- Gauze.
- Lubricating eye ointment.
- Cover glass with diameter 5 mm.
- Large blunt/blunt curved scissors.
- Standard tweezers.
- Hemostat forceps.

- Tray with Styrofoam cooler.
- Gloves.
- Loctite superglue.

2.4 Working Solutions

- 5% Evans Blue solution.
- FITC-dextran.
- Anesthetic—Urethane.

3 Methods

3.1 Anesthetizing the Animals

Before starting any experiments with animals, all the protocols should be approved by the relevant institutional animal care and use committees or relevant ethics committee and researchers should be trained properly to perform surgical procedures. It is also advised to consult with institution veterinarian prior to the start of the animal procedures. Mice are housed in controlled environment in the animal facility and should be allowed to acclimatize for minimum 1 week if brought from an outside facility. Water and food are provided ad libitum. C57BL/6 male mice ranging from 2 to 5 months old are used for imaging experiments usually within the weight range of 22–30 g (*see* **Note 1**).

In this procedure, animals should be weighed initially to determine the anesthesia dose. Anesthesia chosen is normally based on the kind of surgery performed, duration of anesthesia effect, route of administration (i.e., oral or IP), etc. (*see* **Note 2**). Anesthetics should preferably be prepared fresh. Wakefulness of the animal should be ensured by toe-pinch response. Following anesthesia, eye lubricating ointment is applied on the eyes of the animal to prevent dryness. Anesthetics affect the thermoregulation of the animal; hence it is very important to maintain the animal's body temperature by placing it under the lamp or on heat pad throughout the experiment, as this can affect both the reproducibility and mortality rate of the animals (*see* **Notes 3** and **4**). Rectal probe must be used at all times to monitor the body temperature of the animal (*see* **Note 5**).

In this study urethane is used as a terminal anesthesia in mice. Animal should be warmed sufficiently prior to the urethane injection and also post-urethane injection the animal should be continuously placed on the warm pad as most of the anesthetics can cause hypothermia in animals. Urethane is found to be very good for our experiments as it maintains its anesthesia level for up to several hours. **Urethane preparation**: 800 mg of urethane is added to 3 mL of saline to make 4 mL of urethane solution making a final concentration of 200 mg/mL (*see* **Note 6**). **Urethane injection**: Drug is injected at a maximal dose of 2 g/kg to the mice intraperitoneally (IP) (*see* **Note** 7). Typically, for a 30 g mouse, 300 μL of the 200 mg/mL urethane is given. However, only half of the volume is given initially, i.e., 150 μL and the animal is allowed to be anesthetized for 5–10 min, 1/4th, i.e., 70 μL of the total volume is given as second dose and toe-pinch response is assessed and the last 1/4th of the volume is given only if the animal exhibits a pain response on toe-pinch. It usually takes about 20–30 min for the animal to become fully anesthetized.

3.2 Intravenous Tail Vein Injection

Mastering tail vein injection technique is very important prior to performing intravital microscopy (IVM)/Evans Blue dye (EBD) extravasation techniques to assess BBB integrity/permeability. Examples of intravenous injection include injecting under skin into vein using needles, or using catheters, for example in jugular vein. In this case, we chose to perform the intravenous injection into the tail vein, as it is noninvasive in nature compared to inserting catheters into jugular vein. Mouse tail has several small veins; hence, care should be taken to inject the dye or drug into the lateral tail vein. Sterility of the solutions injected is essential and can affect the rate of mortality. Dilation of the vein is extremely helpful and can be done by warming the mouse using a heat lamp or warming heat pad or immersing the tail in warm water (*see* **Note** 8). Typically, 27/30G × ½ in. needles are chosen for tail vein injections in mice. Both the animal and the dye/drug to be injected should be warmed to the body temperature of the animal, which is 37–38 °C. Maximum volume of tail vein injection depends on the size, strain and the vehicle used, for mice it is advisable not to inject beyond 200–250 μL. Care should be taken to remove all the air bubbles from the dye/drug injections given to the animals, as intravenous injection of air bubbles can result in death of the animal.

In this procedure, the needle containing the dye/drug should be inserted into tail at 10–15° angle with the bevel facing up while syringe parallel to the tail. Needle should be inserted into the tail as low as possible and if the injection fails chose a position higher on the tail compared to the previous injection site. Ideal distance of injection is one-third of the tail. Lack of resistance is the best determinant to successful injection into the tail vein. Do not apply back pressure as that can collapse the vein. Restraint devices can also be used for performing tail vein injections based on experience and the level of comfort. Compression of the tail vein proximally is also shown to aid well in tail vein injection.

Apart from tail vein injections, several additional procedures can be performed based on the goal of the experiment including jugular vein cannulations for intravenous drug administrations and carotid artery cannulation for blood pressure and heart rate monitoring, tracheal intubation for artificial ventilation. Femoral and tail vein cannulations can also be done to replace jugular vein injections to inject fluorescent dyes, drugs, antibodies, etc. [21].

3.3 Surgical Procedures

Following anesthesia, animals are set up for surgical procedures. All surgical instruments should be sterilized with cidex or other instrument sterilizers prior to performing any surgical procedure on animals. Animals should be injected subcutaneously with 1 mL of isotonic saline before the surgery to prevent dehydration. Body temperature of the animal should be maintained at 37–38 °C at all times and temperature is monitored using a rectal probe. A thermostatically controlled surgical heating pad may be used preferably.

1. Dorsal side of the head of the animal is shaved using electrical razor.

2. Surgical site on the surface of the head should be wiped with a gauze sponge soaked in 70% ethanol followed by betadine solution.

3. Apply lubricating eye ointment using a sterile Q-tip to prevent the loss of moisture from the eyes during the period of the surgery (see **Note 9**).

4. Animals are injected with 100 μL of fluorescein isothiocyanate (FITC-Dextran-10 kDa; IV) at least 10–15 min prior or post to performing sham or CCI procedures (see **Note 10**). Intravenous injection protocol should be followed as described above.

5. Post FITC injection, animals should be prepped for sham or CCI injury procedures.

6. First, using a forceps gently lift up a small piece of skin at the midline and make a small incision with the help of scissors and remove the skin from top of the skull exposing the sagittal suture, bregma, and lambda.

7. A drop of lidocaine and epinephrine solution may be applied immediately onto the periosteum to avoid excessive bleeding or pain.

8. Periosteum underneath the skull should be gently separated by making an incision using scissors and moving it around using sterile cotton swabs.

9. Perform circular craniotomy window employing a micro drill with a diameter ranging from 3 to 4 mm. Head of the animal should be intermittently irrigated with warmed artificial cerebrospinal fluid (ACSF) or sterile saline to prevent overheating of the brain, as it can hamper with the results (see **Note 11**). The resulting bone flap is removed.

10. Sham animals only receive craniotomy surgery, while CCI induced group receives brain injury following craniotomy procedure.

11. CCI Protocol: Benchmark stereotaxic impactor from Leica is used in this protocol (*see* **Note 12**). Prior to the CCI procedure, dwell time and velocity should be adjusted in the impactor and the right size of the impactor probe tip should be chosen from the settings described in **Note 12** and the position of the impactor shaft should be inclined according to the position of the head of the animal.

    - In this procedure, Extend/Retract toggle switch should be firstly placed in Extend position.

    - Sham induced animal should then be mounted on the stereotaxic frame of the CCI instrument (*see* **Note 13**).

    - Mounting the animal on stereotaxic frame is done in following steps: (1) hold the animal toward one side of the bar and pushing in the other side into the ear canal, (2) open the mouth, put the fixture and tighten.

    - The impact probe should first be centered onto lambda and the anterior/posterior, medial/lateral readings on the digital manipulator should be zeroed.

    - At this point the clip on the end of the lead wire of the contact sensor should be pinned to the skin or the base of the tail or ear of the animal and the impact probe should be lowered onto the brain of the animal, until a beep is heard.

    - Now, the dorsal/ventral reading on the digital manipulator is zeroed and the Retract/Extend toggle switch should be changed back to Retract position.

    - Adjust the dorsal/ventral settings on the digital manipulator to negative values based on the depth of injury chosen. For example, 1 was chosen for mild TBI, -2 for moderate TBI and -3 for severe TBI in this study (*see* **Note 14**).

    - Animal should then be removed from the CCI instrument and placed on a stereotaxic frame and prepped for imaging under the IVM.

12. **Intravital microscopy (IVM)**: It was first described by Wagner in nineteenth century and has been modified over the years. It is a visualization technique used to study cellular and subcellular interactions that occur in live animals and hence is a very useful technique for studying various molecular and physiological changes that take place in animal tissues. It was traditionally used to study microcirculatory changes in post-capillary venules following inflammation or in various in vivo studies employing transparent tissues like cremaster muscle, mesentery, etc.; however, with the help of fluorescently tagged proteins or agents of known molecular weight we can study the cell biology and functions dynamically in opaque tissues like brain and joints. Most commonly used dyes to study various

properties of vasculature include: fluorescent-isothiocyanate (FITC), Rhodamine 6G, and fluorescently tagged antibodies. This method also enables us to study the effect of various pharmacological agents or effect of various genetic manipulations in the in vivo system. Recently, it has been shown to be effective model in studying BBB endothelial cell junctions [20]. The most challenging part of animal studies using IVM is shown to be the preparation of the animal for observation under microscopy following various surgical procedures [20].

Conventional microscope and IVM differ in the way the sample is handled as in vivo conditions animals are thicker than cells on a glass slide hence the working distance between the condenser and the objectives are different. As in vivo imaging employs animals, a large stage is needed so as to hold a variety of additional sensors, tubing, etc. Thermostatic control of animal core and surface temperature is essential. Consequently, need a customized stage most often. Intravital microscopy is mostly associated with imaging and measurement of dynamic events. As we image dynamic events, the data/image storage requirements are also quite high [20, 22].

Set up and imaging using IVM

1. Following sham or CCI injury to animals, the surgical site is cleaned with absorbable gelatin sponges and a small cover glass resembling the size and shape of the removed bone flap is placed on the surgical site and sealed with super glue by gently placing the glue on the outer edge of the surgical site.

2. Microscope should be set up by enclosing it with warm and humidified air or alternatively, the animal, microscope stage and the objective must be warmed by other means, e.g., heating pads warm the animal and the stage while objective heaters are also available for warming up the objective (*see* **Note 15**).

3. The aluminium plate insert under the objective must be cleaned with 70% ethanol.

4. We used inverted microscope with fluoview-1000 laser scanning confocal unit with water immersion Leica 40× W DIC M N2 objective, numerical aperture 0.8, refractive index 1.33 and imaged using a FITC imaging cube with emission wavelength of 525 nm.

5. Open the acquisition software and configure the settings of the intravital microscope. To find the correct focal plane 4× DIC image is taken by moving the objective toward the coverslip until the microcirculation in the brain is clearly visible.

6. To visualize the fluorescence, DIC cube is rotated to FITC cube, this imaging cube has an emission wavelength of 525 nm. Set up all the imaging parameters on the ND acquisition tab.

7. All the vessels are checked initially to evaluate the quality of the preparation and a vessel in which blood is flowing and which is around 20–80 μm in diameter is chosen for imaging in the pial microcirculation.

8. Imaging on intravital microscope can be performed every 20 min up to 2 h. Each spot is recorded for 30 s or a snapshot of the image is taken (*see* Fig. 1).

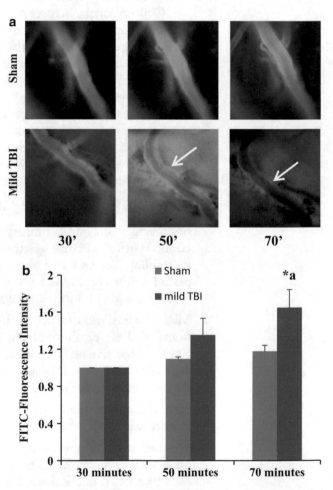

Fig. 1 This figure demonstrates the leakage of fluorescein dextran (FITC-Dextran-10 kDa) from the blood vessel following breach in blood–brain barrier that occurs following Controlled Cortical Impactor-induced traumatic brain injury (TBI) in mouse brain pial vasculature. Animals were injected with FITC-dextran-10 kDa few minutes post TBI injury and were grouped into sham injury (only craniotomy; **a**) and mild TBI injury (2 mm depth with 0.5 m/s velocity and 100 ms contact time; **b**). Following the injury, animals were placed under intravital microscope and the microcirculation in the pial vessels is imaged every 20 min. Representative images are provided in this figure

9. Baseline-integrated optical intensities should be obtained from intravascular and extravascular sites and change in light intensity values should be evaluated. This can be calculated by using the formula: $\Delta I = 1 - (I_i - I_o)/I_i$, where I is the change in light intensity, I_i is the light intensity inside the vessel, and I_o is the light intensity outside the vessel. Experimental values thus obtained should be compared with initial baseline value and expressed as percentage change. This method decreases bias between animals due to red blood cell accumulation and changes in room lighting.

10. After the completion of the procedure, anesthetized mice are euthanized by cervical dislocation.

13. **Evans Blue Extravasation Study**: Evans Blue, also known as T-1824 is an alkaline dye (MW—961 Da) that binds to serum albumin. This dye is extensively used for assessing changes in vascular permeability in various physiological disorders ranging from immunological, inflammatory, cardiovascular, and cancer studies. The dye is traditionally nontoxic in nature and has been used in several concentrations ranging from 1% to 5% and injected at doses ranging from 1 to 5 mg/kg body weight and higher depending on the blood volume of the chosen animal model (*see* **Note 16**). It can be administered through various routes including intravenous, intraperitoneal, and subcutaneous based on the kind of study and the target organ under investigation ([17]; *see* **Note 17**). Restricted exchange of fluids and solutes occurs at the BBB however this condition is altered in a disease condition. Evans Blue works on the principle that under homeostatic conditions, BBB is impermeable to leakage of albumin into the brain, however, this condition is reversed when there is a breach in the integrity of the BBB leading to EBD extravasation into brain tissue, hence, this technique can be used to assess the leakage in BBB. Our studies show that TBI induced animals are shown to have increased leakage of EBD from the blood vessels compared to the non-TBI animals and TBI affected organs show increased blue coloration compared to the nonaffected organs. Changes in vascular permeability using EBD can be either quantified by visualization, e.g., microscopy, or fluorimetric/colorimetric techniques. An advantage of this technique is it can be used to study the variations in BBB permeability in genetically modified mice with different backgrounds [23].

Surgical procedures

1. **Evans Blue Injection into mice**: Prior to Evans Blue injection, animal is anesthetized with urethane as described in the anesthetizing protocol as discussed above. Following anesthesia, each animal is injected with 100 μL of the Evans Blue dye at

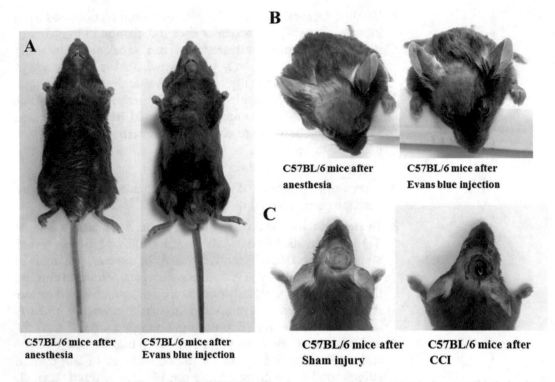

Fig. 2 Sham and Evans Blue-injected C57BL/6 male mice are shown here. Animal after anesthesia is shown on the left (**a**, **b**) while Evans Blue-injected animal is shown on the right side (**a**, **b**). Animal subjected to sham surgery and after subjecting to TBI is shown (**c**)

5% concentration intravenously. Dye can be dissolved in sterile PBS/0.9% saline. Tail vein injection is performed as described in the IV injection protocol section. Evans Blue should be injected 5–30 min before performing the surgery so as to allow the dye to circulate in the system (*see* Fig. 2).

2. **TBI induction**: Following the tail vein injection, animals are inflicted with TBI using the CCI procedure as described above. Precautions in the surgical procedures should be maintained the same as for the intravital protocol.

3. **Organ collection**: On reaching the end point of the study animals are perfused transcardially with ice-cold sterile saline with heparin (1000 U/mL) in order to remove any blood clots (at least 25 mL; *see* **Note 18** for perfusion protocol). Perfusion technique removes the intravascular EB dye, leaving the EB dye in extravascular space of the brain tissue. Animal is decapitated following perfusion and the brain is harvested for performing trichloroacetic acid (TCA) assay. Harvested brains can be used to extract the dye followed by the assayed immediately or stored at −80 °C for performing experiments later. Prior to storing brain tissue, they can be separated into ipsilateral and contralateral halves, weighed and stored in a microcentrifuge tube.

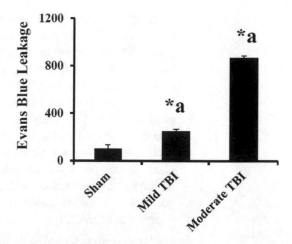

Fig. 3 Effect of controlled Cortical Impactor (CCI)-induced traumatic brain injury (TBI) on Evans Blue (EB) leakage following the breach in the blood–brain barrier. Animals were subjected to sham injury (only craniotomy; $n = 4$), mild TBI injury (1 mm depth with 5 m/s velocity and 300 ms contact time; $n = 5$), and moderate TBI injury (2 mm depth with 5 m/s velocity and 300 ms contact time; $n = 5$). Following sham and TBI procedures brains were harvested and assessed for Evans Blue leakage into brain tissue. The graph on the right hand corner shows a standard curve for Evans Blue concentration in TCA–ethanol with OD measured at 620/680 nm (excitation/emission) with Evans Blue concentrations ranging from 0 to 5000 ng/mL. Data shown here is represented as Mean $\pm$ SEM; $P^* \leq 0.05$. Statistical analysis was performed using one-way ANOVA followed by Bonferroni post hoc test

4. **TCA Assay**: Brain hemispheres can be homogenized using a tissue homogenizer in 50% TCA (w/v). The volume of TCA chosen can be adjusted based on the weight of the brain tissue taken. Homogenate thus obtained will be centrifuged at $6{,}000 \times g$ for 20 min. Supernatant collected will be diluted in three parts of ethanol (1:3, 50% TCA:ethanol). Evans Blue dye can be quantitated either by fluorimetry (620 nm emission/680 nm excitation) or colorimetry (610 or 620 nm) techniques. Samples can be evaluated using corresponding standards ranging from 50 to 1000 ng/mL (*see* **Note 19**; *see* Fig. 3).

4   Notes

1. Vessel permeability also depends on the age and the weight of the animals, hence choosing close birth dates and weights is essential for comparison.

2. Animals can be anesthetized in several ways using a combination of methods or can be used individually including

isoflurane (4% for induction and 1.5–2% for surgery; with 40% O_2 and 60% N_2 via a facemask) followed by ketamine (90–100 mg/kg body weight)/xylazine (5–10 mg/kg dose) intraperitoneal injection, urethane, pentobarbital sodium (40–60 mg/kg, ip), 2,2,2-tribromoethanol (150 mg/kg, IP avertin) per gram of body weight ip (20 μL of 2.5% w/v), pentobarbital (50 mg/kg, ip), halothane(4% and 1.2% for maintenance with 30% oxygen and 69% N_2O), etc. However, each anesthesia has its limitations, e.g., ketamine/xylazine effect wears off after 45 min, while too much inhalation of isoflurane by users is considered a health hazard, and hence, a combination of methods is found to be more beneficial for long-term studies.

3. If a lamp is used to warm the animal, they should be kept at safe working distance to prevent burn injuries.

4. Animals should never be laid directly over the warm pad; there should always be gauze or a cloth separation between the animal and the heating pad. Animals should be continuously monitored for burns.

5. Environmental conditions like temperature and humidity are very important in handling the stress on the animal and all these factors are shown to influence the extent of vascular permeability.

6. Urethane should be heated gently in 37 °C water bath and also vortexed to dissolve any crystals formed.

7. Urethane is carcinogenic and terminal anesthesia, hence care should be taken to use personal protective equipment (PPE) at all times, and also urethane should only be prepared and injected to animals only under chemical hood.

8. Precaution is to be taken while using warm water, as using water at high temperatures can scald the tail tissue leading to a burn injury. Heat pad used should also be maintained strictly at 37–38 °C as low temperatures can lead to hypothermia, while high temperatures lead to loss of thermal regulation and death. Heat lamp should never be placed too close to the animal to prevent burn injury or overheating.

9. Analgesics like buprenorphine (0.01 mg/kg) or dexamethasone (0.2 mg/kg) and carprofeb (5 mg/kg) can be used to prevent pain, swelling, and the inflammatory response in brain prior to or following surgery.

10. FITC-dextran 10 kDa is used in this experiment; however, dextrans of different molecular weights can be studied.

11. Some studies have also employed 1–1.2% agarose in ACSF in order to cover the craniotomy site.

12. Appropriate probe tip among 1, 1.5, 2, and 3 mm is chosen based on the organ of injury and the size of the animal. Travel velocity of the impactor can vary from 1.5–6 m/s with a travel range up to 10.1 mm and dwell time ranging from 0.1 to 9.9 s.

13. Animal's head should be fixed firmly in the stereotaxic frame before impact as it can introduce changes in the injury level. This can be ensured by gently applying pressure on the skull before the impact.

14. The numbers indicate the depth in millimeters into the intact dura. Impactor tip should be wiped clean with sterile alcohol after each impact cleaned/disinfected further with cidex after surgery.

15. Before placing the animal on the microscope stage, the stage needs to be warmed to 37–38 °C. The warm pad and the animal should always be separated by a piece of cloth or gauze. While performing in vivo imaging care must be taken for the animals to tolerate surgery to avoid excessive physiological disturbances that can interfere with the results.

16. Although, EBD is not considered toxic, it is considered as a hazardous irritant and a potential carcinogen; hence it should be used carefully.

17. Duration of time of EBD in system before the experiment depends on the result expected and the kind of barrier under investigation.

18. Perfusion protocol: Place the animal on its back and pin the limbs. Use forceps to lift the skin and body wall. Make an incision laterally and then up each side through the rib cage. Cut through diaphragm and lift the sternum. Pin the loose flap. Needle containing saline should be inserted into the left ventricle and a small incision is made in the right atrium. Animal is perfused for 10 min or until all the blood is perfused out from the right atrium. Decapitate the animal and collect the brain tissue.

19. As the technique is influenced by several factors, it is most advisable to repeat the experiment at least three times.

Acknowledgments

The work presented in this chapter was supported by Scott & White Academic Operations Funds to Dr. Tharakan. The authors would like to apologize to investigators whose works are not cited in this methodological report due to space limitations and the personal perspective with which this chapter has been prepared.

References

1. Deaglio S, Robson SC (2011) Ectonucleoti-dases as regulators of purinergic signaling in thrombosis, inflammation, and immunity. Adv Pharmacol 61:301–332

2. Shen Q, MH W, Yuan SY (2009) Endothelial contractile cytoskeleton and microvascular permeability. Cell Health Cytoskelet 1:43–50

3. Kumar P, Shen Q, Pivetti CD, Lee ES, Wu MH, Yuan SY (2009) Molecular mechanisms of endothelial hyperpermeability: implications in inflammation. Expert Rev Mol Med 11:e19

4. Yuan SY (2002) Protein kinase signaling in the modulation of microvascular permeability. Vasc Pharmacol 39:213–223

5. Pearson WS, Sugerman DE, McGuire LC, Coronado VG (2012) Emergency department visits for traumatic brain injury in older adults in the United States: 2006-08. West J Emerg Med 13:289–293

6. Alluri H, Wiggins-Dohlvik K, Davis ML, Huang JH, Tharakan B (2015) Blood-brain barrier dysfunction following traumatic brain injury. Metab Brain Dis 30(5):1093–1104

7. Thurman DJ, Branche CM, Sniezek JE (1998) The epidemiology of sports-related traumatic brain injuries in the United States: recent developments. J Head Trauma Rehabil 3:1–8

8. Scott BN, Roberts DJ, Robertson HL, Kramer AH, Laupland KB, Ousman SS, Kubes P, Zygun DA (2013) Incidence, prevalence, and occurrence rate of infection among adults hospitalized after traumatic brain injury: study protocol for a systematic review and meta-analysis. Syst Rev 2:68

9. Maas AIR, Stocchetti N, Bullock R (2008) Moderate and severe traumatic brain injury in adults. Lancet Neurol 7:728–741

10. Parikh S, Koch M, Narayan RK (2007) Traumatic brain injury. Int Anesthesiol Clin 45:119–135

11. O'Connor WT, Smyth A, Gilchrist MD (2011) Animal models of traumatic brain injury: a critical evaluation. Pharmacol Ther 130:106–113

12. Kasper C, Yvette C (2015) Traumatic brain injury. In: Kasper C, Conley Y (eds) Annual review of nursing research, vol 33. Springer, New York

13. Gean AD, Fischbein NJ (2010) Head Trauma. Neuroimaging Clin N Am 20:527–556

14. Davidoff G, Morris J, Roth E, Bleiberg J (1985) Closed head injury in spinal cord injured patients: retrospective study of loss of consciousness and post-traumatic amnesia. Arch Phys Med Rehabil 66:41–43

15. Cuccurullo S (2012) Physical medicine and rehabilitation pocket companion. Am J Phys Med Rehabil 91(8):727

16. Xiong Y, Mahmood A, Chopp M (2013) Animal models of traumatic brain injury. Nat Rev Neurosci 14:128–142

17. Nidavani RB, Mahalakshmi AM, Shalawadi M (2014) Vascular permeability and Evans blue dye: a physiological and pharmacological approach. J Appl Pharm Sci 4:106–113

18. Unterberg AW, Stover J, Kress B, Kiening KL (2004) Edema and brain trauma. Neuroscience 129:1021–1029

19. Khan M, Im YB, Shunmugavel A, Gilg AG, Dhindsa RK, Singh AK, Singh IJ (2009) Administration of S-nitroglutathione after traumatic brain injury protects the neurovascular unit and reduces secondary injury in a rat model of controlled cortical impact. J Neuroinflammation 6:32

20. Masedunskas A, Porat-Shliom N, Tora M, Milberg O, Weigert R (2013) Intravital microscopy for imaging subcellular structures in live mice expressing fluorescent proteins. J Vis Exp (79)

21. Marques PE, Oliveira AG, Amaral SS, Nunes-Silva A, Almeida AFS (2012) Intravital microscopy: taking a close look inside the living organisms. Afr J Microbiol Res 6:1603–1614

22. Taqueti VR, Jaffer FA (2013) High-resolution molecular imaging via intravital microscopy: illuminating vascular biology in vivo. Integr Biol (Camb) 5:278–290

23. Radu M, Chernoff J (2013) An in vivo assay to test blood vessel permeability. J Vis Exp (73): e50062

Chapter 5

A Rat Model of Hemorrhagic Shock for Studying Vascular Hyperpermeability

Prince Esiobu and Ed W. Childs

Abstract

Vascular hyperpermeability is one of the known detrimental effects of hemorrhagic shock, which we continually try to understand, minimize, and reverse. Here, we describe induction of hemorrhagic shock in a rat and studying of its effects on vascular permeability, using intravital microscopy. In this protocol, hemorrhagic shock will be induced by withdrawing blood to reduce the mean arterial pressure (MAP) to 40 mmHg for 60 min followed by resuscitation for 60 min. To study the changes in vascular permeability following hemorrhagic shock, the rats will be given FITC-albumin, a fluorescent tracer, intravenously. Following this, the FITC-albumin flux across the vessel will be measured in mesenteric postcapillary venules by determining fluorescent intensity intravascularly and extravascularly under intravital microscopy.

Key words Vascular hyperpermeability, Hemorrhagic shock, Intravital microscopy

1 Introduction

Understanding and being able to effect vascular permeability during shock and critical illness would have a tremendous impact in the medical field. The methods we describe here provide the opportunity to study this phenomenon in a repeatable and efficacious manner in the laboratory. Using this technique, we can change variables (such as with medications, hormones, and genetics) and assess the effects on vascular permeability [1–5].

Hemorrhagic shock (HS) has been implicated in the pathogenesis of multiple organ failure and accounts for 30% of deaths associated with traumatic injury in patients [6]. Vascular hyperpermeability, the excessive leakage of fluids and proteins from the intravascular area to the interstitium, and associated edema occurs as a major clinical complication in trauma following hemorrhagic shock.

Hemorrhagic shock has also been shown by several investigators to induce oxidative stress and reactive oxygen species (ROS) formation and the activation of apoptotic signaling pathways [6–9].

Binu Tharakan (ed.), *Traumatic and Ischemic Injury: Methods and Protocols*, Methods in Molecular Biology, vol. 1717, https://doi.org/10.1007/978-1-4939-7526-6_5, © Springer Science+Business Media, LLC 2018

During reperfusion, a burst of ROS is formed after oxygen is reintroduced into the system. It has been shown that ischemia–reperfusion selectively damages endothelial cells and barriers through a mechanism that involves oxygen-derived free radicals [9, 10]. Thus, oxidative stress due to ROS formation is a major trigger for vascular hyperpermeability and is an important area of research [2]. Furthermore, recent studies have demonstrated that the activation of the apoptotic signaling is a critical to vascular hyperpermeability associated with hemorrhagic shock and the inhibition of this pathway provides protection against vascular hyperpermeability in a rat model of hemorrhagic shock [4].

2 Materials

2.1 Anesthesia

1. 50% urethane solution.
2. 28 Gauge insulin needles.
3. 1 ml syringe.

2.2 Surgery

1. Heated surgical table.
2. Heat lamp.
3. Absorbent underpad.
4. Microdissection curved forceps.
5. Microdissection scissors.
6. 6-0 silk suture.
7. Lab tape.
8. Small hemostat.
9. Normal saline.
10. Heparinized saline.
11. Polyethylene tubing ("PE-50"; 0.58 mm internal diameter).
12. 3-way stop cocks ×3.
13. 23 Gauge 1 in. needles.
14. Micro clip ("bulldog clamp").
15. Large plexiglass plate.
16. Small plexiglass plate.
17. 2 × 2 gauze.
18. Plastic wrap.
19. Rat (Male, 275–325 g).

2.3 Shock

1. Arterial blood pressure monitor.
2. 10 ml syringe ×2.
3. Syringe pump.

2.4 Vascular Permeability	1. Intravital microscope. 2. Intravital microscope camera. 3. Computer with intravital microscopy software. 4. Fluorescein isothiocyanate conjugate—albumin, bovine.

3 Methods

3.1 Anesthesia

1. Secure awake rat in prone position, using your nondominant hand. Make sure that the thighs are exposed.
2. Perform intramuscular (IM) injection of 50% urethane (0.35 ml/100 g of rat weight) into the thigh of a 275–325 g rat (*see* **Note 1**).
3. Place rat back in cage and allow 30 min to 1 h for anesthesia to take full effect.

3.2 Preparation

1. Clip fur on the neck, abdomen, and inner thighs (*see* **Note 2**).
2. Place rate in supine position on heated (43 °C) surgical table covered with an absorbent underpad.
3. Secure all four limbs using tape.
4. Initially, position the head of the rat toward the operator.

3.3 Jugular Vein

1. Grasp the skin of the right neck just superior to the clavicle and cut a 1 cm opening, using scissors.
2. Grasp the fascia and dissect through sharply with scissors.
3. Identify the underlying external jugular vein and bluntly clear surrounding tissue.
4. Pass right angled forceps deep to the jugular vein and isolate it by passing 6-0 silk suture around it.
5. Tie this first suture at the most cephalad portion of the vein. Clamp this suture to the animals jaw under slight tension.
6. Pass another 6-0 silk suture around the most caudal portion of the vein. Make a knot loop but do not tie it down yet.
7. Using the microscissors, make a transverse venotomy on the anterior side of the vessel.
8. Insert saline flushed cannula (bevel side up), connected to a three-way stop cock, into the vessel.
9. Open the three-way stop cock and place below the level of the rat to ensure back flow.
10. Once confirmed, secure the second silk suture over the portion of the cannula within the vessel.
11. Unclamp the first silk suture from the animals jaw and now tie it around the cannula, securing it to the vein (*see* **Note 6**).

3.4 *Carotid Artery*

1. Identify the midline of the neck. On either side, there will be the sternohyoid muscle. Just lateral, there will be the sternocephalic muscle. Bluntly separate these muscles, exposing the carotid artery, lateral to the trachea.

2. Hook the carotid artery with right angled forceps and bluntly dissect the loose surrounding tissue.

3. Pass right angled forceps deep to the vessel and isolate it by passing 6-0 silk suture around it.

4. Tie this first suture at the most cephalad portion of the artery. Clamp this suture to the animals jaw under slight tension.

5. Place a bulldog clamp at the inferior portion of the vessel.

6. Pass another 6-0 silk suture around the most caudal portion of the artery. Make a knot loop but do not tie it down yet.

7. Using scissors, make a transverse arteriotomy on the anterior portion of the vessel.

8. Insert heparinized saline flushed cannula (bevel side up), connected to a three-way stop cock, into the vessel.

9. Secure cannula to the chin of the rat, using index finger of nondominant hand, then remove the bulldog clamp.

10. While still securing cannula, use other hand to quickly open and close stop cock to ensure bleeding.

11. Flush cannula with heparinized saline.

12. Now tighten inferior knot; release superior knot from clamp on the chin and tie it around the cannula (*see* **Note 6**).

3.5 *Femoral Artery*

1. Now turn the head of the rat away from the operator.

2. Identify the depression three quarters of the distance from the pubic tubercle to the inner border of the leg.

3. Grasp the skin with forceps and cut a 1.5 cm circular opening, using scissors.

4. Sharply cut through fascia using scissors.

5. Identify femoral vessels above the take-off of the epigastric vessels.

6. Use right angled forceps to hook the artery and pass a 6-0 silk suture around vessel; Tie this suture as caudally as possible just above the level of the epigastric vessels (*see* **Note 3**).

7. Place a bulldog clamp at the superior portion of the artery.

8. Pass another 6-0 silk suture just below the clamp; make a loop but do not tighten it.

9. Make a transverse arteriotomy and insert the stretched heparinized saline flushed cannula.

10. Loosely secure the superior suture.

11. Remove the clamp and flash the stop cock to ensure back bleeding, then flush with heparinized saline.

12. Secure the suture (*see* **Note 6**).

3.6 Mesentery

1. Identify the midpoint between the xyphoid process and pubic tubercle; make a midline longitudinal incision extending an inch in either direction.

2. Incise the fascia sharply at the visible white line in the midline.

3. Untape the limbs at this point and very gently reposition the rat onto a plexiglass plate on its left side; you can pick the rat up by its limbs to accomplish this repositioning.

4. Place a smaller plexiglass in front of the abdominal side of the rat; apply saline to the surface of this glass.

5. Massage the abdominal contents out by placing thumbs on right lateral abdominal wall and applying pressure on the posterior surface of the spine.

6. Using a cotton-tip applicator, splay out the bowel and its mesentery on the smaller plexiglass.

7. Identify a capillary of choice within the clear mesentery (*see* **Note 4**).

8. Using the cotton-tip applicator, carefully return all other segments of bowel into the abdominal cavity.

9. Place multiple normal saline soaked 2×2 gauze sponges over the bowel, while leaving the mesentery and chosen capillary exposed (*see* **Note 5**).

10. Cut an appropriately sized piece of clear plastic wrap and lay it over the mesentery and small bowel. It should lie in contact with the mesentery (*see* **Note 6**).

3.7 Preshock

1. After a 30 min recovery period, place the rat (still on plexiglass) on the microscope with the capillary in view through the scope (Fig. 1).

2. Inject FITC-albumin (50 mg/kg) at this time through the jugular catheter (*see* **Note 7**) (Fig. 1b).

3. Place an anal temperature probe. Temperature should be between 36 and 38 °C.

4. Connect the femoral artery stop cock to a blood pressure transducer.

5. Connect a 10 ml syringe (previously flushed with heparinized saline) to the carotid artery catheter.

6. Connect the jugular catheter to a syringe pump filled with normal saline. Start infusing at a rate of 3 ml/h.

7. At this point, take and save a preshock microscopy image (*see* **Note 6**) (Fig. 1a, b).

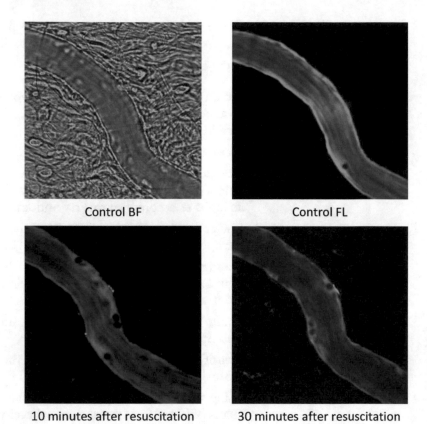

Control BF   Control FL

10 minutes after resuscitation   30 minutes after resuscitation

Fig. 1 The images shown are of mesenteric post capillary venules of sham control (bright-field and fluorescent), hemorrhagic shock for 1 h MAP 40 mmHg followed by 10 and 30 min after resuscitation

3.8 Shock

1. Using the empty syringe connected to the carotid catheter, draw out blood from the rat until the mean arterial pressure (MAP) is down to 40 mmHg.

2. During the shock period, maintain the MAP around 40 mmHg (38–42 mmHg) by drawing more blood out or returning blood as is necessary.

3. Perform this for the desired period of time. We typically shock the rats for 60 min.

4. After this period, return the shed blood to the rat over a 5 min period.

5. Now supplement with normal saline as needed to maintain MAP above 90 mmHg.

6. Take microscopy images at desired intervals. We typically take images at 10 min intervals post shock (Fig. 2) (*see* **Note 6**).

3.9 Intravital Microscopy

1. Open your images with your intravital microscope software.

2. On the pre shock control (Tc) image, analyze the optical intensity at two locations both inside and outside of the vessel. Average each and create an out–in ratio.

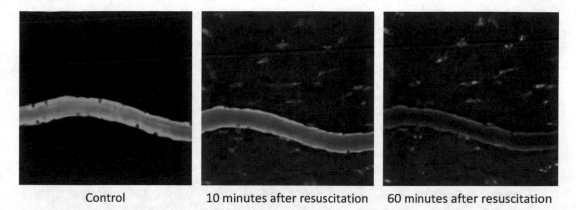

| Control | 10 minutes after resuscitation | 60 minutes after resuscitation |

Fig. 2 Intravital microscopy images of rat mesenteric post capillary venules of sham control and hemorrhagic shock 1 h MAP 40 mmHg followed by 10 and 60 min after resuscitation

3. Repeat this for the rest of your time (Tx) points during the shock period, using the same locations along the vessel as the control image.

4. Obtain a ratio (Tx:Tc) for each of your time points.

5. Plot and compare your ratios in desired fashion.

4   Notes

1. For faster anesthetic effect, you can give half of the anesthetic dose into each thigh.

2. The rats should be fasted for 18 h prior to the procedure.

3. Due to the smaller caliber of the femoral artery, stretch the PE-50 cannula at the tip, causing it to narrow.

4. The capillaries in the mesentery are almost invisible to the untrained eye. You will be looking for a very small vessel. The ideal capillary is 25–30 μm. Look carefully.

5. Keep the mesentery moist at all times with warm normal saline. The overlying plastic wrap reduces evaporation.

6. The rat should be normothermic (~37 °C). Heat lamps may be necessary during surgery and during shock.

7. Turn lights off when handling the FITC-albumin.

References

1. Childs E et al (1999) Leukocyte adherence and sequestration following hemorrhagic shock and total ischemia in rats. Shock 11:248–252

2. Childs EW, Udobi KF, Wood JG, Hunter FA, Smalley DM, Cheung LY (2002) *In vivo* visualization of reactive oxidants and leukocyte-endothelial adherence following hemorrhagic shock. Shock 8:423–427

3. Tharakan B et al (2010) (−)-Deprenyl inhibits vascular hyperpermeability after hemorrhagic shock. Shock 33:56–63

4. Childs E et al (2007) Apoptotic signaling induces hyperpermeability following hemorrhagic shock. Am J Physiol Heart Circ Physiol 292:H3179–H3189

5. Childs E et al (2010) 17β-estradiol mediated protection against vascular leak after hemorrhagic shock: role of estrogen receptors and apoptotic signaling. Shock 34:229–235

6. Murao Y, Hata M, Ohnishi K, Okuchi K, Nakajima Y, Hiasa Y, Junger WG, Hoyt DB, Ohnishi T (2003) Hypertonic saline resuscitation reduces apoptosis and tissue damage of the small intestine in a mouse model of hemorrhagic shock. Shock 20:23–28

7. Childs EW, Udobi KF, Hunter FA, Dhevan V (2005) Evidence of transcellular albumin transport after hemorrhagic shock. Shock 23:565–570

8. Davidson MT, Deitch EA, Lu Q, Hasko G, Abungu B, Nemeth ZH, Zaets SB, Gaspers LD, Thomas AP, DZ X (2004) Trauma-hemorrhagic shock mesenteric lymph induces endothelial apoptosis that involves both caspase-dependent and caspase-independent mechanisms. Ann Surg 240:123–131

9. Therade-Matharan S, Laemmel E, Duranteau J, Vicaut E (2004) Reoxygenation after hypoxia and glucose depletion causes reactive oxygen species production by mitochondria in HUVEC. Am J Physiol Regul Integr Comp Physiol 287:R1037–R1043

10. Savoye G, Tamion F, Richard V, Varin R, Thuillez C (2005) Hemorrhagic shock resuscitation affects early and selective mesenteric artery endothelial function through a free radical-dependent mechanism. Shock 23:411–416

Assessment of Cardiovascular Function and Microvascular Permeability in a Conscious Rat Model of Alcohol Intoxication Combined with Hemorrhagic Shock and Resuscitation

Travis M. Doggett, Jared J. Tur, Natascha G. Alves, Sarah Y. Yuan, Srinivas M. Tipparaju, and Jerome W. Breslin

Abstract

Hypotension, cardiac depression, and elevated microvascular permeability are known problems that complicate resuscitation of patients following traumatic injury, particularly those who are also intoxicated from alcohol consumption. A conscious rat model of combined alcohol intoxication and hemorrhagic shock has been used to study the hemodynamic mechanisms involved. Here, we describe using this model to study microvascular leakage and cardiac electrical activity.

Key words Hemorrhage, Shock, Alcohol, Microcirculation, Microvascular permeability, Electrocardiogram

1 Introduction

Acute alcohol intoxication is a significant health problem in the USA and contributes to an increased risk of traumatic injury, accounting for the majority of alcohol-related cases in emergency rooms [1, 2]. Binge drinking is a nationwide trend involving at least 38 million adults in the USA, which by the most recent estimates produced a $223.5 billion cost burden for the USA [3]. Nearly 40% of injured patients admitted to the ER have a blood alcohol concentration greater than 80 mg/dL [2, 4–6]. Intoxicated patients present a complex, altered physiologic state associated with increased injury severity [4, 6–9]. In addition, these patients typically require a significantly greater frequency of interventions in the hospital such as endotracheal intubations, placement of intracranial monitoring devices, and greater use of diagnostic peritoneal lavage [7]. Compared to their sober counterparts, alcohol-intoxicated

Binu Tharakan (ed.), *Traumatic and Ischemic Injury: Methods and Protocols*, Methods in Molecular Biology, vol. 1717,
https://doi.org/10.1007/978-1-4939-7526-6_6, © Springer Science+Business Media, LLC 2018

trauma patients have a twofold increase in infectious complications, development of pulmonary failure, acute respiratory distress syndrome (ARDS) secondary to sepsis, and an overall increase in ARDS-related mortality [10].

One particularly problematic outcome of acute alcohol intoxication is aggravated hemodynamic instability following hemorrhage [9]. The dramatic loss of circulating fluids elicits a massive baroreceptor reflex, reducing splanchnic blood flow and causing ischemic injury that disrupts integrity of the gut wall. This leads to a systemic inflammatory response, including systemic microvascular leakage [11–14]. The resulting edema complicates fluid resuscitation and can lead to sepsis, ARDS, and abdominal compartment syndrome [15].

A conscious rat model of alcohol intoxication and hemorrhagic shock was developed by Molina and colleagues to test whether an impairments in the baroreceptor reflex or associated neural, endocrine, and metabolic mechanisms that control vascular tone may be responsible for this increased hypotension [14, 16–18]. Their findings showed that: (1) significantly less blood from alcohol-intoxicated rats is required to produce the same hypotensive pressure during a fixed-pressure hemorrhage; and (2) alcohol-intoxicated rats are significantly more hypotensive following equal volumes of removed blood [14, 18]. In this same model, alcohol-intoxicated rats had depressed blood bicarbonate levels, lower P_{O2}, elevated P_{CO2}, and lower plasma glucose and lactate compared to control rats [18]. Alcohol intoxication also causes altered tissue and plasma cytokine responses during hemorrhagic shock [18]. The elevations in plasma levels of epinephrine, norepinephrine, and arginine vasopressin in response to fixed-pressure hemorrhage were all significantly blunted in alcohol intoxicated rats compared to control rats [17]. However, their findings also showed that sympathetic control of blood pressure and acetylcholine-induced vasodilation were not impaired by alcohol intoxication [19, 20]. Intracerebroventricular administration of choline, a precursor of the preganglionic neurotransmitter acetylcholine, did not improve hemodynamic and neuroendocrine counter-responses to hemorrhagic shock in alcohol-intoxicated rats [19]. Furthermore, studies using isolated aorta and mesenteric arteries of intoxicated and control rats that underwent experimental hemorrhage showed that the alcohol-induced impairment of hemodynamic regulation is not due to a decrease in responsiveness of blood vessels to vasoconstrictors like phenylephrine, or vasodilators like acetylcholine [20].

Such findings have led us to investigate alternative mechanisms. One such potential mechanism is a reduction in central fluid volume due to elevated microvascular permeability. To approach this, we have combined the aforementioned rat alcohol intoxication model with intravital microscopy. Our results showed that alcohol intoxication increases microvascular leakage in the mesentery, which could potentially contribute to alcohol-induced hypotension [21]. In the current paper, details of the combination of these intravital microscopy methods with acute alcohol-intoxication (AAI) and hemorrhagic shock and resuscitation (HSR) in rats are described. A second potential mechanism is direct action on the electrical activity of the heart [22–24]. ECG is a minimally invasive procedure that allows recording of cardiac electrical function. Lead II format specifically identifies left ventricular function with higher specificity. The use of lead II ECG recording following combined AAI and HSR in the rat is also described in this protocol.

2 Materials

Catalog numbers are for reference to instruments in the authors' laboratories. Similar instruments based on cost and preference can be substituted.

2.1 Catheter Preparation

1. Polyethylene 50 tubing (Becton Dickinson).
2. Silastic I.D. 0.51 mm × O.D. 0.94 mm (Dow Corning).
3. Bunsen burner.
4. Scissors.
5. Paperclip loop.
6. Ring stand and clamp.

2.2 Catheter Implantation Surgery

1. Fine scissors (Fine Science Tools (FST) 14958-11).
2. Tissue forceps (FST 11021-14).
3. Halsey needle holders (FST 12501-13).
4. Extrafine forceps (FST 11152-10).
5. Vannas scissors (FST 15000-00).
6. Bulldog clamp (FST 18054-28).
7. Micro dissecting forceps (Roboz RS-5069).
8. Ultrafine hemostats (FST 13021-12).
9. Bead sterilizer (FST 18000-50).
10. Beveled trocar I.D. 3 mm × O.D. 4 mm.
11. Hemostats wrapped with Tygon tubing.
12. Betadine scrub (Purdue Products).
13. Chlorhexidine gluconate scrub (Baxter).
14. Alcohol.
15. 0.9% sodium chloride USP (Hospira).
16. 10 cc syringes (Becton Dickinson).
17. 23 gauge needles (Becton Dickinson).

18. Gauze sponges (Fisher).

19. Ethicon FS-2 suture and needle (Johnson & Johnson).

20. Ethicon 4-0 silk, black, braided suture without needle (Johnson & Johnson).

21. Perry style 42 gloves (Ansell).

22. Lighter.

23. Masking tape

24. Silastic-tipped vascular catheters.

25. Gastric catheters.

26. Isoflurane USP and vaporizer (Henry Schein and Harvard apparatus).

27. Cotton-tipped applicators.

28. Akwa tears ophthalmic ointment (Akorn).

29. Fur trimmers (Oster).

2.3 Initial Preparation Procedures

1. Mouse-sized cage.

2. Digital scale (Mettler Toledo).

3. 1 cc syringe with plunger removed and luer tip cut off (Becton Dickinson).

4. 3 cc syringe (Becton Dickinson).

5. 23 gauge needles (Becton Dickinson).

6. 0.9% sodium chloride USP (Hospira).

7. Fine scissors (Fine Science Tools (FST) 14958-11).

8. Hemostats wrapped with Tygon tubing.

9. Masking tape.

10. Lighter.

11. Heated water bath (Fisher).

12. Lactated Ringers USP (Hospira).

2.4 Fixed-Pressure Hemorrhage

1. Heated water-pad and circulating heated water pump (Gaymar).

2. Software and hardware for measuring MAP: LabChart Pro v7, PowerLab 4/35, Quad Bridge Amp (ADInstruments).

3. Pressure transducer (ADInstruments).

4. 3 cc syringes (Becton Dickinson).

5. Heparin-coated 3 cc syringes.

6. 23 gauge needles (Becton Dickinson).

7. 3-way stopcocks (Braun).

8. 1.5 mL centrifuge tubes (Eppendorf).

9. Refrigerated centrifuge (Eppendorf).

10. Cryo-vials (Corning).

11. 0.9% sodium chloride USP.

12. Water-filled beaker.

2.5 Resuscitation

1. Software and hardware for measuring MAP: LabChart Pro v7, PowerLab 4/35, Quad Bridge Amp (ADInstruments).

2. Pressure transducers (ADInstruments).

3. Warm Lactated Ringers USP (Hospira).

4. 10 cc syringes (Becton Dickinson).

5. 30 cc syringes (Becton Dickinson).

6. 20 gauge needles (Becton Dickinson).

7. Polyethylene 90 tubing (Becton Dickinson).

8. Hemostats wrapped with Tygon tubing.

9. Fine scissors (Fine Science Tools (FST) 14958-11).

10. Syringe pump (KD Scientific).

11. Lighter.

2.6 Intravital Microcopy

1. 2 × 2 Gauze sponges (Fisher).

2. 3 × 3 Gauze sponges (Fisher).

3. Betadine scrub (Purdue Products).

4. Chlorhexidine gluconate scrub (Baxter).

5. Alcohol.

6. Fine scissors (Fine Science Tools (FST) 14958-11).

7. Tissue forceps (FST 11021-14).

8. Hemostats wrapped with Tygon tubing.

9. Bead sterilizer (FST 18000-50).

10. Lactated Ringers USP (Hospira).

11. Heating plate and temperature controller (World Precision Instruments).

12. Isoflurane USP and vaporizer (Henry Schein and Harvard apparatus).

13. FITC-albumin (Sigma-Aldrich).

14. Syringe pump (Fisher).

15. Peristaltic pump (Cole-Palmer).

16. Water-bathed heating coil (Radnoti).

17. Circulating water bath and pump (Thermo).

18. Software and hardware for measuring MAP: Lab Chart Pro v7, Power Lab 4/35, Quad Bridge Amp (AD Instruments).

19. Pressure transducer (AD Instruments).

20. Upright fluorescent microscope (Nikon E600).

21. Mercury lamp and power source (Nikon).

22. Fur trimmers (Oster).

23. Cotton-tipped applicators.

24. Euthasol—pentobarbital sodium and phenytoin sodium (Virbac).

2.7 Surface Lead Electrocardiogram (ECG) Recording

1. Isoflurane USP and vaporizer (Henry Schein and Harvard apparatus).

2. Three lead monopolar needle electrodes (29 gauge).

3. Software and hardware for measure ECG: Lab Chart Pro v7, Power Lab 8/35, Animal Bio Amp (ADInstruments).

4. Heating plate and temperature controller along with rectal probe (World Precision Instruments).

3 Methods

The fixed-pressure hemorrhage/resuscitation (HSR) protocol requires implantation of polyethylene catheters into the carotid artery, jugular vein, and an indwelling gastric catheter. The carotid catheter is used for withdrawing blood during the HSR phase of the protocol as well as monitoring, in real-time, mean arterial pressure (MAP) during HSR and intravital microscopy. The jugular catheter allows for the infusion of resuscitative fluids during the HSR and FITC-conjugated albumin tracer during intravital microscopy. The indwelling gastric catheter allows for administration of alcohol or water prior to experimentation. This method was chosen over oral gavage to minimize stress on the rats, which can occur with handling. Isovolumic water serves as a control for distension of the stomach. This was chosen over isovolumic, isocaloric dextrose solution, which we observed causes changes in lymphatic pumping [25].

A fixed-pressure HSR protocol was chosen to ensure the rats received the same degree of injury severity. Studies have shown rats that with fixed-volume hemorrhage, the alcohol-intoxicated rats are significantly more hypotensive [14, 18]. In addition, this protocol is performed in conscious, unrestrained rats, preventing any compounding effects on hemodynamics caused by anesthetics, and stress-induced increases in sympathetic output caused by restraints. This method also better simulates alcohol intoxication and traumatic injury in the human population.

Intravital microscopy is a powerful tool for determining microvascular permeability in vivo [26–28]. For this model, the mesenteric microcirculation is a key site of interest due to the ischemia-reperfusion injury on the gut caused by HSR. FITC-albumin is typically used as a tracer for extravasation of plasma macromolecules. Elevated fluorescence in the extravascular spaces over time indicates increased microvascular leakage. To determine that increases in leakage are due to changes in diffusive permeability, diameters of the arterioles feeding the local microvascular bed are also monitored. If these diameters increase, the leakage is likely due to increased filtration, while if these diameters do not change, or even decrease, there is likely an increase in permeability of microvascular endothelium. In addition, rapid acquisition of brightfield images in sequence can be obtained for measurement of red blood cell velocity as well as leukocyte rolling and adhesion in postcapillary venules.

Surface lead ECG recordings are a minimally invasive procedure that demonstrates the electrical activity of the heart. In 2005 the FDA issued that well-controlled electrocardiograms (ECG) trials are required to define cardiac risk of new therapies [29]. Indeed ECGs have been a powerful tool for cardiovascular health and its ability to acutely sense toxic effects of chemicals and or drugs [30]. The QT and QTc duration correspond to the repolarization phase of the cardiac action potential [31, 32]. The QRS complex and the P wave are also studied utilizing the bipolar limb leads in lead I, II, and III positioning. Lead II positioning provides a clear image of the QRS complex and ST segment the ECG analysis. The ECG protocol is performed while the rats are anesthetized with 2% Isoflurane inhalation. Heart rate is allowed to stabilize (5 min) followed by lead II ECG signal acquisition.

All animal protocols should be performed in strict accordance government regulations. In this USA this includes the US Animal Welfare Act, US Public Health Service Policy on the Humane Care and Use of Laboratory Animals, and the Guide for the Care and Use of Laboratory Animals. All animal experiments must be performed with the approval of the Institutional Animal Care and Use Committee in the USA or equivalent regulatory committee in other countries. Aseptic technique is required for all surgical procedures including catheter implantation and the laparotomy for intravital microscopy. All surgical instruments to be used must be sterilized prior to surgical procedures by autoclave and again just before use with a bead-sterilizer. Areas of skin where incisions will be made for surgery must be properly cleaned with antiseptic cleansers. All fluids and drugs used should be US Pharmacopeia (USP)-grade. All efforts should be made to minimize pain, including providing analgesics after surgery.

3.1 Catheter Preparation

1. Catheters used for this protocol are hand-made from commercially available Intramedic polyethylene 50 (PE50) tubing from Becton Dickinson.

2. Using a thick metal paper clip, create a circular loop and mount it to a ring stand. Heat the loop with a Bunsen burner until it glows red then remove the flame (*see* **Note 1**).

3. Allow a few seconds for the loop to stop glowing then feed about 1 in. of PE50 through the loop.
   The tubing will begin to melt and contract on itself to create a bubbled section of tubing.
   This bubble will serve as an anchor point for sutures to prevent the catheters from pulling out (*see* **Note 2**).
   This step is applies to both vascular and gastric catheters.

4. Measure off about 40 cm from the bubbled end and cut to produce one vascular catheter. For gastric catheters, measure 46 cm from the bubbled end and cut.

5. Cut off excess PE50 tubing from the bubble-side of the catheter leaving about 0.5 cm from the end to the bubble (*see* **Note 3**).

6. Catheters implanted into the blood vessels have an additional strip of silastic tubing attached to the catheter.
   Precut 5 cm long strips of silastic and slide onto the bubble-end of the catheter. Continue sliding the silastic onto the catheter until it covers the bubble (*see* **Notes 4** and **5**).

7. Trim the silastic down by cutting the end to 3 cm in length from the end of the bubble (*see* **Note 6**).

8. Each rat will receive two vascular catheters (one for the carotid artery and one for the jugular vein) and one gastric catheter (*see* **Note 7**).

3.2 Catheter Implantation Surgery

1. In order to perform the fixed-pressure HSR protocol, catheters are implanted into the left common carotid artery and right jugular vein 4–5 days prior to the experiment. An indwelling gastric catheter is also implanted for administration of alcohol or water to the stomach.

2. For anesthesia, isoflurane is a good choice; however, other anesthetics such as ketamine/xylazine will also work. To anesthetize with isoflurane, place the rat into the induction chamber. Set the vaporizer to 4%, open the USP-grade oxygen gas cylinder, and set the flow rate to 1 liter (L) per minute (min).

3. Once the rat is sufficiently anesthetized, remove from the induction chamber and place with the dorsal side and insert the nose into the nosepiece that will deliver additional isoflurane. Reduce the isoflurane to about 2.5% (*see* **Note 8**).

4. To prevent the rat's eyes from drying out, apply Akwa tears ophthalmic ointment using cotton-tipped applicators.

5. Using commercially available hair trimmers, shave about a 2 × 2 cm area of fur from the dorsal nape of the neck starting about 5 mm inferior form the ears. This is the location for the incision through which the catheters will be routed out.

6. Place the rat ventral side up and shave the fur from the ventral neck area. Again, about a 2×2 cm area depending on the size of the rat should be shaved of fur.

7. Clean the shaved areas three times: once with Chlorhexidine gluconate surgical scrub, once with alcohol, and once with Betadine scrub.

8. Before starting the surgery, the catheters to be used should be flushed and filled with saline using a 10 cc syringe. Ensure there is no air bubbles within the catheters.

9. With the rat in a dorsally recumbent position, make a 1 cm incision in the ventral neck along the medial long axis.

10. Carefully isolate the left common carotid artery. Ultrafine hemostats are good for separating tissue and isolating the vessel.

    Consultation of a resource such as *Anatomy & Dissection of the Rat 3RD ed.* by Walker and Homberger for the location of the carotid is recommended.

11. Once the carotid is isolated, suture the brain side of the vessel closed with an 8 cm long strip of 4-0 suture. Tie off with two simple suture knots.

12. Place another 8 cm long strip of 4-0 suture under the isolated carotid artery on the heart side.
    Do not tie off this suture yet.

13. To facilitate insertion of the catheter into the artery, a small drop of lidocaine can be administered to the isolated vessel.

14. Place a small bulldog clamp on the heart side of the vessel (*see* **Note 9**).

15. Using Vannas scissors, create a small incision in the carotid artery (*see* **Note 10**).

16. Using micro dissecting forceps to hold open the incision, feed one of the silastic-tipped catheters into the vessel with extrafine forceps.

17. Remove the bulldog as the catheter is inserted into the vessel taking care to keep a firm hold on the catheter (*see* **Note 11**).
    Continue to feed the catheter into the vessel to the bubble (*see* **Notes 12** and **13**).

18. Replace the bulldog clamp to hold the catheter in place. Secure the catheter inside the vessel with 4-0 suture placed earlier on both sides of the bubble (*see* **Note 14**).
    Two simple suture knots are sufficient on to anchor the catheter.

19. Remove the bulldog clamp and check for adequate blood flow. Flush a small amount of saline to clear any blood from the catheter and thermally seal the end with a lighter (*see* **Note 15**).

20. Repeat **steps 8–13** for the right jugular vein (*see* **Notes 16** and **17**).

21. Turn the rat over onto its ventral side. Create a small 0.5 cm incision on the midline of the dorsal nape of the neck about 1 cm inferior from the ears.

22. Using a trocar, route the catheters subcutaneously to and out the incision of the dorsal nape of the neck.

23. Using Ethicon FS-2 suture and needle or similar, close the ventral neck incision with a continuous, locking suture or individual simple suture knots.

24. Lay the rat onto its right side. Using your index finger, feel for the bottom of the rib cage. Make a 2 cm long incision perpendicular to the long axis of the rat (dorsal to ventral) into the skin.

    The incision should be about 3.5 cm away from the bottom of the ribcage (about the width of the index and middle finger).

25. Separate the skin, in the immediate area of the incision, from the muscle layer using extrafine hemostats.

26. Make a second incision slightly smaller in length into the muscular layer and open the peritoneal cavity.

27. Locate and externalize the stomach through the incision. The stomach should be just under the liver.

    Be sure to only grab the stomach by the muscular layer. Do not grab by the transparent tissue.

28. Using Ethicon FS-2 suture and needle begin a purse-string knot within the muscular layer of the stomach. Do not tighten the suture.

29. Perforate the stomach with small diameter scissors such as Vannas scissors and insert a non-silastic tipped catheter bubble-end first into the perforation. Finish tightening and secure the purse-string knot.

30. Secure the gastric catheter in place through the top of the muscular layer incision with a single simple suture then finish closing the incision with a continuous locking suture.

31. Using a trocar, route the gastric catheter subcutaneously to and out the incision of the dorsal nape of the neck. The catheters should be marked with sharpie or marker in unique ways to identify the catheters.

32. Close the incision in the skin with a continuous locking suture with Ethicon FS-2 suture and needle (*see* **Note 18**).

33. Place the rat back on its ventral side. Pull the catheters taut and secure them to the incision on the dorsal nape of the neck with simple sutures using Ethicon FS-2 suture and needle.

(a) It helps to secure the carotid catheter to the left side of the incision and the jugular to the right side of the incision and the gastric catheter in between (*see* **Note 19**).

34. Once the catheters are secure with suture, close the remaining open incision with simple suture knots.

35. To prevent the rat from gnawing on the exposed catheters, coil them around your index finger and wrap with masking tape. If secured properly, the taped catheters will stand up perpendicular from the body out of the rat's reach.

36. Allow the rat 4–5 days to recover before experimentation.

3.3 Initial Preparation Procedures

1. There are several companies that provide software and hardware capable of measuring mean arterial blood pressure (MAP) in real time. Our lab has used Lab Chart Pro v7 software and with the Power Lab 4/35 and Quad Bridge Amp provided by AD Instruments.

2. On the day of the experiment, rats to be used should be weighed and placed into mouse sized-cages. Weighing the rat is important, as this will be used to estimate the total blood volume (TBV) of the rat, calculated as the volume equivalent to 7% of body weight (BW).

3. Unwrap catheters and feed them through a 1 cc syringe with the plunger removed and luer end cut off. This is to prevent the rat from manipulating the catheters and chewing on them.

4. Locate the carotid artery line. Cut off the end to ensure flow of arterial blood. If no blood flows from the line, flush with saline using a 3 cc syringe and check again. Remove the syringe and check for blood flow from the line. Once blood is able to flow through the catheter, flush with saline to prevent clotting and thermally seal with a lighter.

5. Locate the jugular line and clamp it with the hemostat. It is important to clamp the line first with the jugular because once the line is opened it will pull air in. Too much air will cause an air embolism and can cause death. Cut the end of the line and flush it with saline from the 3 cc syringe to ensure fluids can be infused. Thermally seal the catheter with a lighter.

6. Feed the catheters through a hole made in the top of the mouse cage and secure with tape. Ensure to leave enough slack in the catheters between the rat and the top of the cage to allow for unrestricted movement.

7. Bring the cage(s) to the HSR equipment. Cages may be placed onto a heated water-pad to help keep the rats warm during the HSR protocol if desired.

8. The pressure transducers for the Power Lab 4/35 are connected to the carotid catheter via a 23 G needle. Connect

the carotid catheter to the pressure transducer and flush the line slightly with saline provided by a 10 cc syringe reservoir atop the transducer rig (*see* **Note 20**).

9. Open Lab Chart Pro v7 on the computer. Press the "start" button in the software to begin pressure recordings. If the pulse pressure (PP) is narrow check for kinks created in the catheters and flush the line. This should correct and restore PP. Perform this as needed through the entire HSR protocol.

10. Because the rats are unrestrained and the catheters are secured to the top of the cage, constant observation of the state of the catheters must be maintained. The rat moving around leads to twists, coils, and even kinks that are further compounded by the fact there are three catheters.

3.4 Alcohol Administration and Fixed-Pressure Hemorrhage

1. Record 60 min of baseline MAP. This will allow the rat time to relax. This permits any sympathetic discharge due to handling, weighing, and transfer from the vivarium to dissipate.

2. Following baseline MAP readings, 2.5-g/kg dose of alcohol or isovolumic water is administered to the rat via the gastric catheter.

    (a) MAP readings are recorded for an additional 30 min for sufficient absorption of alcohol.

3. During the hemorrhage phase, warm a 500 mL bag of USP-grade lactated Ringer's (LR) in a water bath set to 37 °C. This will be used for the resuscitation phase of the protocol.

4. The hemorrhage phase of the protocol is initiated 30 min after administration of alcohol or water (*see* **Note 21**). Blood is withdrawn to achieve a fixed-pressure hemorrhage between 40–60 mm Hg for 60 min.

    Blood withdrawal is achieved using a 3 cc syringe and connecting it to a 3-way stopcock attached to the transducer rig.

5. A large bolus of blood is removed within the first minute, typically between 4.5 and 6.0 cc of blood depending on the BW of the rat (330–365 g range). Remove more if needed to achieve 40–60 mm Hg in this initial phase of the hemorrhage (*see* **Note 22**).

6. Once the desired map is achieved, flush the catheter slightly with saline to prevent blood clotting in the line. Do not flush too much saline back during the hemorrhage phase. Just enough to clear most of the blood from the catheter (*see* **Note 23**).

    The catheter must be flushed with saline with each subsequent blood withdrawal.

7. Log the volume of blood withdrawn. We estimate TBV from the rat's BW (7% of BW). We set 50% of TBV as the upper limit of the volume of blood that can be safely removed.

   Generally, 44 ± 2% of TBV from water-treated and 36 ± 1% of TBV for alcohol-treated rats is removed.

8. Once 40–60 mm Hg MAP is achieved, maintain this range by withdrawing blood as necessary for the remaining 60 min.

   Usually withdrawing 0.5 cc at a time is sufficient to maintain the hemorrhage MAP.

9. Blood samples may be kept from various time points during the protocol. A fresh heparinized 3 cc syringe should be used to withdraw the blood for these samples.

   The samples should be spun in a refrigerated centrifuge at 2000 rpm (425 × g) and 4 °C to obtain plasma.

   Snap freeze the plasma in liquid nitrogen and store at −20 °C.

10. Following 1 h of hemorrhage the resuscitation phase of the HSR protocol begins.

3.5 Resuscitation

1. Resuscitation consists of a 40% total blood volume removed (TBR) bolus and a 2× TBR infusion for 1 h of warm (37 °C) LR administered via the jugular catheter. Remember to clamp the jugular prior to cutting off the thermally sealed end of the catheter to prevent air from being pulled in.

2. Following completion of the hemorrhage, determine the final TBR for each rat and withdraw the appropriate volume of LR prewarmed to 37 °C into a 10 cc syringe (for the 40% TBR bolus) and a 30 cc syringe (for the 2× TBR infusion).

3. Using a 20-gauge needle and PE90 tubing, connect to the jugular catheter and administer the 40% TBR bolus of warm LR over 1 min.

4. Clamp the jugular line following administration of the bolus and remove the 20 gauge needle from the 10 cc syringe and place it onto the 30 cc syringe containing the infusion fluids (*see* **Note 24**).

5. Place the syringe into a syringe pump. Set the appropriate syringe settings based on the brand used, the total volume to be infused, and the rate to infuse that volume over 60 min (*see* **Note 25**). Start the syringe pump.

6. Continue to record MAP using Lab Chart and Power Lab during the resuscitation phase of the protocol.

7. Once the 2× TBR infusion concludes the HSR protocol is complete. Thermally seal catheters and proceed to either the intravital microscopy (Subheading 3.6) or the ECG (Subheading 3.7) phase of the protocol.

3.6 Intravital Microscopy

1. Immediately following completion of resuscitation, transfer the rat to the surgery station for intravital microscopy preparation.

2. Anesthetize the rat with isoflurane (other methods for anesthesia may be used if desired) (4% induction and 2.5–1.5% maintenance).

3. Once sufficiently anesthetized place the rat in a dorsally recumbent position onto a heating pad set to 37 °C connected to an adaptive temperature controller. Shave away the fur from the ventral abdominal region using rodent hair clippers.

4. Clean the shaved area three times with 3 × 3 gauze sponge: once with Chlorhexidine gluconate surgical scrub, once with alcohol, and once with Betadine scrub.

5. Perform a midline laparotomy in the ventral abdominal skin. The incision length should be only between 2.5 and 3.0 cm long (*see* **Note 26**).

6. Using single 2 × 2 gauze sponges, insert one into the left side of the incision and one into the right side. This helps to hold in the abdominal contents and prevent additional intestine from spilling out onto the stage.

7. Using cotton-tipped applicators carefully begin to remove the mesentery from the abdominal cavity (*see* **Notes 27** and **28**).

8. Turn the rat onto its side and move the heating pad and rat to the intravital microscopy stage (*see* **Note 29**).

9. Apply some warm LR to the stage.

10. Carefully splay the mesentery flat over the stage using cotton-tipped applicators and only handling by the small intestinal walls. Apply some warm LR to the mesentery to prevent drying out (*see* **Note 30**).

11. Once the mesentery is splayed out onto the stage, transfer the entire setup (rat, stage, and heating pad) to the microscope.

12. Begin a superfusion drip of LR using a peristaltic pump to about 10–12 drips per minute. LR should be pumped through a water-bathed heating coil heated to 37 °C by a circulating water bath/pump. Remove excess LR from the stage reservoir by a vacuum line.

13. Begin administration of FITC-albumin to the jugular catheter. FITC-albumin administration consists of a bolus of 1 mg/10 kg in 1 mL LR over 1–2 min followed by continuous infusion of 0.15 mg/kg/min. The vehicle for FITC-albumin is warm LR.

14. For the continuous infusion, set the syringe pump to one half the rate used during resuscitation. Wait 10–15 min before imaging to allow for uniform distribution of FITC-albumin throughout the circulation.

15. During the 10–15 min incubation period, connect the carotid catheter to a pressure transducer to monitor MAP during. The same equipment used during HSR can be used.

16. Begin imaging of the mesenteric microcirculation using an upright fluorescent microscope. Our lab captures images for 30 min at intervals of 0 (baseline), 10, 20, and 30 min.

17. Fluorescent stills are captured with a 4× objective followed by a 30 s of brightfield images (time-lapse movie) with a 10× objective. Two separate areas of microcirculation are imaged with at least one containing a lymphatic vessel.

18. Images are analyzed using ImageJ open-source software. Fluorescent images are used for quantifying extravasation of FITC-albumin by measuring the integrated optical intensity (IOI) of the extravascular regions of the postcapillary venules. The mean 30-min IOI is calculated and plotted and statistics performed. All data is normalized to the 0 min baseline.

19. Brightfield movies are for assessing leukocyte rolling and adhesion, lymphatic pumping, and arteriolar diameter measurements.

3.7 Surface Lead Electrocardiogram (ECG) Recording

1. Transfer the rat to the ECG station for ECG preparation.

2. Anesthetize the rat with isoflurane (other methods may be used if desired) (4% induction and 2.5–1.5% maintenance), or keep the rat anesthetized if ECG is recorded after intravital microscopy.

3. Once sufficiently anesthetized, place the rat in a dorsally recumbent position onto a heating pad set to 37 °C connected to an adaptive temperature controller. Insert the rectal probe to establish rat temperature and record.

4. Remove the hair just below the paws of the rat by shaving and if possible hair removal cream. Clean the limbs three times with 3 × 3 gauze sponge with alcohol.

5. For lead II configuration place the positive lead (red wire) into the left hind paw, the negative lead (black wire) into the right front paw, and finally the reference lead (green wire) into the right hind paw. Alternatively if available bio potential "snap on" leads can be used in place of needle leads.

6. The Animal Bio Amp for the Power Lab (4/35 or 8/35) should be connected to the ECG needle leads.

7. Open Lab Chart Pro on the computer. Press the "start" button in the software to begin ECG recordings. If the ECG trace is noisy check the placement of the leads, pause the software and readjust the leads if necessary. Care should be taken to place the leads just under the skin but not deep into the skeletal muscle, which can confound some of the electrical signaling. Perform this as needed through the entire ECG protocol.

8. Record 5 min of baseline ECG to ensure a stable heart rate. This will allow the heart rate to become stable while under anesthesia. Begin acquiring ECG signals in lead II position, pause the software and reposition the leads to I and then III. Lead I position is moving the positive lead (red wire) to the right front paw; and lead III is moving the positive lead (red wire) to the right hind paw and moving the reference lead (green wire) to the left hind paw.

4 Notes

1. The metal loop and ring stand. This loop is simply made by taking a paper clip and straightening out one end and forming a loop on the other. The looped end is heated with a bunsen burner. One end of the PE50 tubing is fed through and the heat from the loop melts the tubing to form a bubble(*see* Fig. 1).

2. PE50 tubing with a bubble formed at one end at one end after holding it within a metal loop heated by a Bunsen burner. The bubble serves as an anchor point for securing the catheter inside of the vessel or stomach. This method is used when making both vascular and gastric catheters (*see* Fig. 2).

3. Trimmed bubble-end of the catheter. Once the bubble is formed, the end of the catheter should be trimmed to about 0.5 cm from the end of the bubble. Both vessel and gastric catheters undergo this step. Gastric catheters are complete and ready for use after completion of this step (*see* Fig. 3).

4. To facilitate sliding silastic tubing over the bubble of the catheter, the silastic strips should be soaked in chloroform for 10 min.

   This helps make the silastic more compliant and easier to get over the bubble of the catheter.

   Perform in a well-ventilated room or under a fume hood.

Fig. 1 Metal loop and ring stand assembly for catheter preparation

Fig. 2 PE50 tubing with a bubble formed at one end at one end after holding it within a metal loop heated by a Bunsen burner

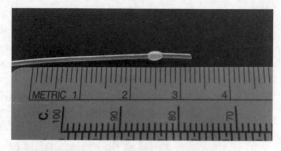

Fig. 3 PE50 tubing catheter with bubble after trimming

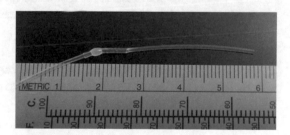

Fig. 4 Vascular catheter with silastic tubing attached

5. Vascular catheters are the same as gastric catheters with the addition of a strip of silastic tubing on the bubble-end. Silastic is less rigid and more compliant than polyethylene. It facilitates ease of cannulation of the vessel and protects against vessel damage that otherwise may be caused by the rigid polyethylene tubing. The vascular catheter with silastic tubing attached is shown (*see* Fig. 4).

6. After the silastic tubing is placed on the vascular catheter, the silastic is trimmed to about 3 cm from the end of the bubble. This is the average length from where the incision is made in the carotid artery that provides the best signal when recording MAP. The end of the silastic should be trimmed with a slight bevel to facilitate insertion of the catheter into the vessel.

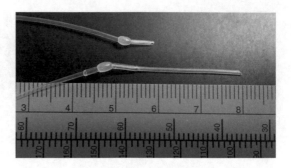

Fig. 5 The gastric (*top*) and vascular (*bottom*) catheters are shown, demonstrating the differences in length of the silastic tubing. Note the slightly beveled tip in the silastic of the vascular catheter

7. Each rat will receive two vascular catheters (one in the carotid artery and one in the jugular vein) and one gastric catheter. The vascular and gastric catheters are initially made the same way as described in **step 3** of catheter preparation. Vascular catheters have an additional strip of silastic tubing. The gastric (top) and vascular (bottom) catheters are shown, demonstrating the differences in length of the silastic tubing. Note the slightly beveled tip in the silastic of the vascular catheter (*see* Fig. 5).

8. There is no set percentage value for isoflurane during maintenance. Each rat responds differently. Continually monitor the breathing rate during surgery and adjust the percentage up or down as needed.

9. Placement of a bulldog clamp is absolutely essential. If no clamp is placed onto the carotid before an incision is made, the rat will begin to hemorrhage and the vessel will be quickly lost under the blood.

10. Take care not to completely cut through the artery. The remaining blood in the isolated portion of the artery will signal a successful incision.

11. Blood pressure will push the catheter out if not held firmly inside the vessel.

12. Blood will begin flowing into the catheter. If not flushed quickly with saline, the blood will begin to coagulate.

    To prevent this the catheter should be clamped with a Tygon-wrapped hemostat.

13. When clamping catheters, it is important to use Tygon-wrapped hemostats.

    Clamping catheters with hemostats without a rubber wrapping or coating will crimp the PE50 tubing.

    These hemostats can be easily made by sliding a strip of Tygon tubing onto the ends of a hemostat.

14. Using the bulldog to hold the catheter in place while suturing is recommended. The rat's blood pressure will push the catheter out if it is not secured in place by the bulldog clamp.

15. It is extremely important to ensure the carotid catheter is well sealed.

    A small break in the seal will allow enough of a pressure differential to fill the catheter with blood. By the end of the recovery period, this blood will be coagulated and the catheter will be useless.

16. The jugular vein is extremely thin-walled and easy to tear when isolating with ultrafine hemostats.

17. The jugular catheter will pull in air if not clamped. Ensure it is clamped before incision to prevent air emboli.

18. When suturing the skin incision, be careful not to pierce the gastric catheter with the needle.

19. Take care not to pierce the catheters especially the carotid and jugular with the needle.

20. The pressure transducer rig is comprised of transducer housing and several components.

    One end of the transducer has a 10 cc syringe reservoir with 0.9% saline USP for flushing blood from the carotid after each withdrawal.

    The other end has a three-way stopcock and 23-gauge needle for connection to the carotid line.

21. Total baseline MAP readings are 90 min. Sixty minutes are recorded before and 30 min after alcohol or water is administered intragastrically.

22. It is extremely important to get the rats MAP down to 40–60 mm Hg as quickly as possible. Otherwise the rat's sympathetic compensatory mechanisms will keep MAP close to normal until the rat decompensates and death occurs.

23. Blood withdrawal will cause an interruption in the MAP recording. In addition, flushing the catheter with saline will introduce artifacts in the recording. We plot our MAP recordings using samples or 10 s at intervals every 5 min.

    Take precautions not to withdraw blood or flush the line during these time points or whichever you choose to plot.

24. Clamp the line when changing the bolus syringe for the infusion syringe. This will prevent introduction of air bubbles inside the jugular line that could cause air emboli.

25. Most commercially available syringe pumps are preprogramed with the most common brands, composition, and volume syringes. However, if the syringe is not preprogrammed, it

will be necessary to know the diameter of the syringe and program it into the syringe pump manually.

26. Generally, it is helpful to make the incision just off the midline to the side with which the rat will be laying. This will help the mesentery lay flat on the stage for microscopy imaging.

27. The microcirculation within the mesentery is delicate. To prevent hemorrhaging of microvessels, the mesentery should only be handled and manipulated by the intestinal walls of the small intestines.

28. At this point, when externalizing the mesentery, take time to untwist it. This will make splaying the mesentery flat, as well as moving new areas onto and off the stage much easier.

29. There are commercially available intravital microscopy stages or one can be engineered from materials. Our stage is a simple design consisting of a Plexiglas base, 3 cm culture dish serving as stage, and walled-off super fusion fluid reservoir.

30. The warm LR often facilitates intestinal motility, including the migrating myoelectric complex. These are rhythmic, migrating muscle contractions along the length of the intestinal wall that can interfere with microscopy of the mesenteric microcirculation. To control for this, use 2×2 gauze sponges soaked with LR and place them onto the small intestinal wall. This weighs down the intestine and helps to hold the imaging area in place.

References

1. Howard RJ, Slesinger PA, Davies DL et al (2011) Alcohol-binding sites in distinct brain proteins: the quest for atomic level resolution. Alcohol Clin Exp Res 35:1561–1573

2. Whitaker AM, Sulzer JK, Molina PE (2011) Augmented central nitric oxide production inhibits vasopressin release during hemorrhage in acute alcohol-intoxicated rodents. Am j Physiol Regul Inter Comp Phsyiol 301:R1529–R1539

3. CDC (2012) Vital signs: binge drinking prevalence, frequency, and intensity among adults - United States, 2010. MMWR Morb Mortal Wkly Rep 61:14–19

4. Vonghia L, Leggio L, Ferrulli A et al (2008) Acute alcohol intoxication. Eur J Intern Med 19:561–567

5. Cherpitel CJ, Bond J, Ye Y et al (2003) Alcohol-related injury in the ER: a cross-national meta-analysis from the Emergency Room Collaborative Alcohol Analysis Project (ERCAAP). J Stud Alcohol 64:641–649

6. Madan AK, Yu K, Beech DJ (1999) Alcohol and drug use in victims of life-threatening trauma. J Trauma 47:568–571

7. Jurkovich GJ, Rivara FP, Gurney JG et al (1992) Effects of alcohol intoxication on the initial assessment of trauma patients. Ann Emerg Med 21:704–708

8. Hadfield RJ, Mercer M, Parr MJ (2001) Alcohol and drug abuse in trauma. Resuscitation 48:25–36

9. Shih HC, Hu SC, Yang CC et al (2003) Alcohol intoxication increases morbidity in drivers involved in motor vehicle accidents. Am J Emerg Med 21:91–94

10. Maier RV (2001) Ethanol abuse and the trauma patient. Surg Infect 2:133–141. discussion 141–144

11. Carden DL, Granger DN (2000) Pathophysiology of ischaemia-reperfusion injury. J Pathol 190:255–266

12. Childs EW, Tharakan B, Hunter FA et al (2007) Apoptotic signaling induces hyperpermeability following hemorrhagic shock. Am J Physiol Heart Circ Physiol 292: H3179–H3189

13. Childs EW, Tharakan B, Byrge N et al (2008) Angiopoietin-1 inhibits intrinsic apoptotic

signaling and vascular hyperpermeability following hemorrhagic shock. Am J Physiol Heart Circ Physiol 294:H2285–H2295

14. Mathis KW, Zambell K, Olubadewo JO et al (2006) Altered hemodynamic counter-regulation to hemorrhage by acute moderate alcohol intoxication. Shock 26:55–61

15. Kumar P, Shen Q, Pivetti CD et al (2009) Molecular mechanisms of endothelial hyperpermeability: implications in inflammation. Expert Rev Mol Med 11:e19

16. Greiffenstein P, Mathis KW, Stouwe CV et al (2007) Alcohol binge before trauma/hemorrhage impairs integrity of host defense mechanisms during recovery. Alcohol Clin Exp Res 31:704–715

17. Molina PE, Zambell KL, Norenberg K et al (2004) Consequences of alcohol-induced early dysregulation of responses to trauma/hemorrhage. Alcohol 33:217–227

18. Phelan H, Stahls P, Hunt J et al (2002) Impact of alcohol intoxication on hemodynamic, metabolic, and cytokine responses to hemorrhagic shock. J Trauma 52:675–682

19. Mathis KW, Molina PE (2009) Transient central cholinergic activation enhances sympathetic nervous system activity but does not improve hemorrhage-induced hypotension in alcohol-intoxicated rodents. Shock 32:410–415

20. Molina MF, Whitaker A, Molina PE et al (2009) Alcohol does not modulate the augmented acetylcholine-induced vasodilatory response in hemorrhaged rodents. Shock 32:601–607

21. Doggett TM, Breslin JW (2014) Acute alcohol intoxication-induced microvascular leakage. Alcohol Clin Exp Res 38:2414–2426

22. Zhang Y, Post WS, Dalal D et al (2011) Coffee, alcohol, smoking, physical activity and QT interval duration: results from the Third National Health and Nutrition Examination Survey. PLoS One 6:e17584

23. Aasebo W, Erikssen J, Jonsbu J et al (2007) ECG changes in patients with acute ethanol intoxication. Scand Cardiovasc J 41:79–84

24. Lorsheyd A, de Lange DW, Hijmering ML et al (2005) PR and QTc interval prolongation on the electrocardiogram after binge drinking in healthy individuals. Neth J Med 63:59–63

25. Souza-Smith FM, Kurtz KM, Molina PE et al (2010) Adaptation of mesenteric collecting lymphatic pump function following acute alcohol intoxication. Microcirculation 17:514–524

26. Duran WN, Sanchez FA, Breslin JW (2008) Microcirculatory exchange function. In: Tuma RF, Duran WN, Ley K (eds) Handbook of physiology: microcirculation, 2nd edn. Academic Press—Elsevier, San Diego, CA, pp 81–124

27. Breslin JW, Wu MH, Guo M et al (2008) Toll-like receptor 4 contributes to microvascular inflammation and barrier dysfunction in thermal injury. Shock 29:349–355

28. Hatakeyama T, Pappas PJ, Hobson RW 2nd et al (2006) Endothelial nitric oxide synthase regulates microvascular hyperpermeability in vivo. J Physiol 574:275–281

29. Morganroth J (2007) Cardiac repolarization and the safety of new drugs defined by electrocardiography. Clin Pharmacol Ther 81:108–113

30. Farraj AK, Hazari MS, Cascio WE (2011) The utility of the small rodent electrocardiogram in toxicology. Toxicol Sci 121:11–30

31. Couderc JP (2009) Measurement and regulation of cardiac ventricular repolarization: from the QT interval to repolarization morphology. Philos Trans A Math Phys Eng Sci 367:1283–1299

32. Varro A, Baczko I (2011) Cardiac ventricular repolarization reserve: a principle for understanding drug-related proarrhythmic risk. Br J Pharmacol 164:14–36

Chapter 7

Intracerebral Hemorrhage in Mice

Damon Klebe, Loretta Iniaghe, Sherrefa Burchell, Cesar Reis, Onat Akyol, Jiping Tang, and John H. Zhang

Abstract

Intracerebral hemorrhage is the most devastating stroke subtype with high rates of mortality and morbidity. Furthermore, no clinically approved treatment exists that effectively increases survival or improves quality of life for survivors. Effective modeling is necessary to elucidate the pathophysiological mechanisms of intracerebral hemorrhage and evaluate potential therapeutic approaches. Rodent models are most utilized because of their cost-effectiveness, and because rodent brain development and structures are well documented. Herein, we describe two intracerebral hemorrhage mouse models: the autologous blood double-injection and collagenase infusion models.

Key words Autologous blood, Collagenase, Intracerebral hemorrhage, Stroke, Mouse, Animal model

1 Introduction

Intracerebral hemorrhages constitute 10–15% of stroke subtypes and have very poor prognoses, since the mortality rate is between 30% and 50%, and at least 75% of survivors are incapable of independent living after 1 year [1–3]. Decompressive craniotomy, controlled ventilations for reducing intracranial pressure, and lowering blood pressure are the only viable emergency interventions [4]. Intracerebral hemorrhage prevalence increased by 18% over the last decade, and incidence is expected to continue increasing due to a large, aging population, who are most vulnerable [2]. Minimal advancements have been made for intracerebral hemorrhage clinical management, and no pharmacotherapy has been clinically approved that effectively increases survival or improves the quality of life for survivors [5]. In order to facilitate discovering pathophysiological mechanisms of intracerebral hemorrhage and help advance novel therapeutic approaches, an effective animal model best demonstrating the clinical situation is a necessity.

Intracerebral hemorrhage injury progression commences with mechanical pressure and shear force applied to brain tissue from the

Binu Tharakan (ed.), *Traumatic and Ischemic Injury: Methods and Protocols*, Methods in Molecular Biology, vol. 1717,
https://doi.org/10.1007/978-1-4939-7526-6_7, © Springer Science+Business Media, LLC 2018

hematoma mass effect. Force and pressure imposed onto glia and neurons result in consequent secretion of excitotoxic neurotransmitters, which lead to cytotoxic edema, apoptosis, and necrosis [6, 7]. Although bleeding terminates soon after the initial injury, rebleeding and subsequent hematoma expansion is observed in approximately 30% of clinical intracerebral hemorrhage cases [8–10]. Additionally, blood components activate microglia, resident central nervous system macrophages, and systemic immune cells infiltrate brain tissue, all of which secrete proinflammatory cytokines, proteases, and oxidative species that further damage brain tissue and augment blood–brain barrier disruption [11–13].

Both the autologous blood double-injection and the collagenase infusion mouse models mimic specific aspects within intracerebral hemorrhage pathophysiology. Both models also have inherent strengths and weaknesses that need to be taken into consideration when designing studies. The autologous blood double-injection model best mimics the hematoma mass effect, yet the model does not involve actual rupturing of major blood vessels. The collagenase infusion model best mimics the bleeding, rebleeding, and consequent hematoma expansion, yet bleeding is slow, diffuse, and ruptures only small vessels and capillary beds; a major vessel is not the main source of the initial bleed. Furthermore, collagenase exacerbates the immune response, making it a confounding factor when studying inflammation after intracerebral hemorrhage [14, 15]. Herein, we describe how to perform both mouse models, which involve stereotaxic guided surgical injections directly into the brain.

2   Materials

2.1 Surgical Equipment

1. Stereotactic frame with mouse adaptor (Stoelting, Wood Dale, IL, USA).
2. Infusion pump (Harvard Apparatus, Holliston, MA, USA).
3. Rodent surgical heating pad.
4. Mouse restraint cones.
5. 1 mL 26 gauge syringes.
6. Size 10 scalpel.
7. 3–1/2 in. iris scissors.
8. 6 in. curved forceps.
9. Cotton tipped applicators.
10. A microdrill with 1 mm diameter drill bits (CellPoint Scientific, Gaithersburg, MD, USA).
11. 10 µL Hamilton syringe for collagenase/250 µL Hamilton syringe for blood (Hamilton Company, Reno, NV, USA).

12. 26 gauge needle.

13. Nonheparinized capillary tubes for blood collection.

14. Timer.

15. Bone wax.

16. 6 in. clamp hemostat.

17. 6 in. straight forceps.

18. 3-0 silk sutures.

19. Induction chamber and recovery chamber.

20. Heating blanket.

2.2 Reagents

1. Ketamine–Xylazine anesthesia cocktail (*see* **Note 1**).

2. Atropine (*see* **Note 2**).

3. 70% isopropyl alcohol.

4. Betadine.

5. Ophthalmic lubricant.

6. Type VII-S Collagenase (*see* **Note 3**).

7. Buprenorphine (*see* **Note 4**).

8. Saline.

3 Methods

3.1 Preinjection
Surgical Procedures

1. Wear appropriate personal protective equipment (gloves, masks, head cover) when handling animals at all times. Mice typically used are 8 weeks old, male, CD1 mice, weighing approximately 30–35 g.

2. Weigh the mouse then calculate the amount of Ketamine–Xylazine cocktail (1.5 mL/kg), Atropine (0.5 mL/kg), and saline (10 mL/kg) needed. Fill 1 mL 26 gauge syringes with the calculated volumes needed, one syringe for each solution injected.

3. Place the mouse in the cone restraint and position the animal supine using the nondominant hand with the posterior end slightly elevated. Inject anesthesia by inserting the syringe into the lower abdominal quadrant at a 30° angle to the skin. Always aspirate first to ensure the needle has not penetrated into any viscera. If blood, urine, or gastrointestinal contents fill into the tip of the syringe, then the solution is contaminated and the needle must be removed and discarded. The procedure must be performed again with a new syringe and solution. Inject the anesthesia cocktail, keep the needle in place for 3 s, then remove and inspect the injection site to make sure that all the

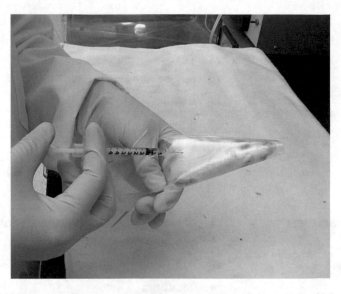

Fig. 1 Mouse placed in a cone restraint and receiving an intraperitoneal injection. Angle the mouse downward and insert the needle, bevel faced up, into one of the lower abdominal quadrants at a 30° angle to the skin. Always aspirate first to make sure that no blood, tissue, or gastrointestinal contents enters the syringe, which confirms the needle has not perforated any visceral tissue and is properly located in the intraperitoneal cavity

solution was injected properly. Repeat the same procedure for Atropine (*see* Fig. 1).

4. Place the mouse in an induction cage, which should be protected from light and noise to avoid stimulating the animal while the anesthesia takes effect. Test toe-pinch response to ensure anesthesia is sufficiently deep. If appropriate anesthesia is achieved, muscle tone will be relaxed and the pedal withdraw reflex from pressure applied to paws (toe-pinch response) will be absent.

5. When anesthesia is sufficiently deep, slightly wet hair on the scalp with tap water and cut the hair using iris scissors. Perform this procedure away from the surgical station to keep the station relatively clean.

6. Using cotton-tipped applicators, apply 70% isopropyl alcohol followed by betadine to the shaved scalp, which will disinfect the surgical site and reduce chances for infection.

7. Moisten the eyes with ophthalmic lubricant to prevent drying during surgery.

8. Use the scalpel to make a vertical midline incision on the scalp (1–1.5 cm) (*see* **Note 5**). Remove the soft tissue covering the skull by blunt dissecting using cotton-tipped applicators. Make sure that bregma is clearly visible.

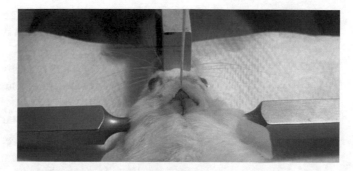

Fig. 2 Mouse mounted to the stereotaxic frame. The head should be completely horizontal and parallel to the plane of the table. The head is secured properly if it does not move when the animal breathes and if slight pressure can be applied to the skull without moving the head

9. Place the animal prone onto the heating pad, which should be programmed to maintain a core body temperature around 37 °C. To mount the head, fix the animal on the stereotactic frame in a prone position, using the front teeth to secure the mouth to the bite plate of the frame, making sure to pull the tongue to the side using the curved forceps to keep the airway clear. Tighten the frame above the nose, making sure the clamp is not too tight to restrict breathing. Place the ear pins securely either in the ears or on the sides of the skull and fasten shut. Adjust the bite plate and ear pins such that the head is flat and level. Make sure that the head is secured such that it does not move when the animal breathes or when slight pressure is applied to the top of the skull (*see* Fig. 2).

10. Aspirate the Hamilton syringe with distilled water and place in the injection pump on the stereotactic frame, with the tip at bregma and the bevel facing the midline. Bregma is point 0 (*see* **Note 6**).

11. Move the needle to the following coordinates relative to bregma, which will be the injection site: X (right lateral) $= 2.2$ mm; Y (rostral) $= 0.2$ mm.

12. Use the drill to make a 1 mm cranial burr hole at the X, Y position. (*see* **Note 7**).

3.2 Autologous Blood Double-Injection

1. Program the infusion pump to deliver 30 μL in a 250 μL Hamilton syringe at a rate of 5 μL/min.

2. Remove the Hamilton syringe from the infusion pump apparatus. Release the animal from the stereotactic frame and place in a supine position, allowing easy access to the central tail artery. Disinfect the tail with 70% isopropyl alcohol, which also helps dilate the central tail artery, and allow the alcohol to air dry.

Using a 26 gauge needle, puncture the artery and collect a minimum of 30 μL of blood in a nonheparinized capillary tube (*see* **Note 8**).

3. Quickly transfer the collected blood to the 250 μL Hamilton syringe (*see* **Note 9**) and make sure that no air bubbles are in the syringe. Place the syringe back on the microinjection pump, making sure the needle bevel is facing toward the midline.

4. Remount the mouse on the stereotactic frame, and insert the needle into the cranial burr hole until the tip is just below the calvarium. This is point 0 depth-wise (*see* **Note 10**). Further lower the needle to 3.0 mm below the dura.

5. Infuse 5 μL of blood, then pause. Lower the needle an additional 0.7 mm, wait for 5 min, and then infuse 25 μL.

6. Leave the needle in place for 10 min and withdraw at a rate of 1 mm/min to prevent backflow.

3.3 Collagenase Infusion

1. Program the infusion pump to deliver 0.5 μL in a 10 μL Hamilton syringe at a rate of 0.133 μL/min.

2. Remove the Hamilton syringe from the infusion pump apparatus. Prepare the collagenase by removing from the freezer and thawing. Aspirate the needle (10 μL Hamilton syringe) with collagenase three or four times, to ensure the entire column is coated with collagenase. Finally, aspirate 8–10 μL of collagenase, remove any air bubbles, and place the needle back onto the frame, such that the bevel is facing the midline (*see* **Note 11**).

3. Insert the needle into the burr hole until the tip is just below the calvarium. Advance the needle ventrally until it reaches a depth of 3.7 mm.

4. Infuse 0.5 μL of the collagenase solution (0.075 Units) (*see* **Note 12**). Leave the needle in place for an additional 5 min, and then retract at a rate of 1 mm/min to prevent backflow.

5. For sham surgery, repeat the procedure for collagenase infusion, only replace collagenase solution with phosphate buffered saline.

3.4 Post-Injection Surgical Procedures and Recovery

1. Using a cotton-tipped applicator, apply bone wax to the burr hole immediately after the needle is removed, making sure the hole is completely plugged.

2. Using the hemostat, forceps, and 3-0 silk suture, suture the incision shut by preferably making a running suture.

3. Inject saline subcutaneously into the flanks of the hind limbs. Inject buprenorphine (0.03 mg/kg) subcutaneously as needed to alleviate pain.

4. Remove the animal from the stereotactic frame and place into the recovery cage, which is placed on top of the heating blanket to maintain a temperature near 37 °C within the cage.

5. Once the mouse is awake, alert, and moving, place it back into its original cage.

4 Notes

1. Prepare anesthesia cocktail by mixing Ketamine (100 mg/mL) with Xylazine (20 mg/mL) in a 2:1 ratio. For example, to make 1.5 mL anesthesia cocktail, mix 1 mL Ketamine with 0.5 mL Xylazine.

2. Atropine (0.4 mg/mL) is given to reduce secretions in the airway, allowing the animal to maintain easier breathing while under anesthesia.

3. Prepare collagenase solution by dissolving sterile filtered, chromatography purified type VII collagenase from *Clostridium histolyticum* (Sigma-Aldrich, St. Louis, MO, USA) into ice-cold, sterile filtered phosphate buffered saline to create a 0.15 Unit/μL solution. Aliquot into 0.2 mL microcentrifuge tubes at 20 μL/tube and store in −20 °C freezer protected from light. Collagenase preparation needs to be performed on ice and very rapidly to minimize degradation.

4. Buprenorphine is given after surgery to alleviate any pain and discomfort the animal experiences from the procedure.

5. An incision that is too short can make the surgery cumbersome (i.e., finding and drilling at the X, Y coordinates) as there is only a small surgical window, while too long of an incision may cause damage to surrounding tissue.

6. Bregma is an anatomical point at the intersection of the coronal and sagittal sutures on the superior middle portion of the skull. Bregma is the main reference point for stereotactic brain surgery (*see* Fig. 3). After aspirating the syringe with distilled water, ensure that all the water is removed to prevent unwanted dilution of the collagenase. Diluted collagenase can result in smaller hematomas, which will affect the injury severity and create variations in the neurobehavioral tests.

7. Drill gently, and avoid damaging tissues below the dura, which can result in excessive bleeding. Placing the drill at a 45° angle can help to reduce the possibility of damaging the underlying tissue. After drilling, check to ensure that the burr hole is in the correct location by lowering the needle to the top of the hole.

8. Dipping the tail in lukewarm water will also help dilate the central artery, making it easier to puncture and draw blood.

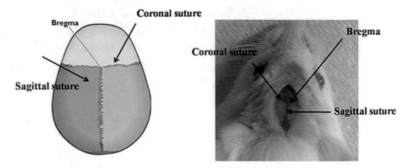

Fig. 3 Diagram and picture of bregma. Bregma is the intersection point of the coronal suture and sagittal suture. Bregma serves as the point of origin in stereotactic surgeries

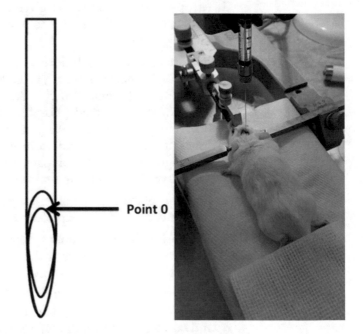

Fig. 4 Point 0 at the top of the hole in the needle bevel. The bevel of the needle is facing toward the midline. When inserting the needle through the burr hole into the brain, the point of origin for depth is when point 0 is flush with the top of the skull

9. Working fast while collecting and transferring blood into syringe prevents blood clotting and ensures smooth infusion of blood into the striatum. The autologous blood model is a double-injection procedure because a small amount of blood is injected first above the main target site to create a plug in the needle track that significantly decreases the chance of backflow.

10. Point 0 for depth is when the very top of the hole in the needle bevel is flush with the top of the skull when inserting the needle into the burr hole (*see* Fig. 4).

11. The needle may be flushed with 2–3 µL of collagenase between animals to remove any residual collagenase from the previous animal. This ensures that none of the collagenase being infused has been at room temperature for an extended period, thus potentially reducing its potency.

12. Check that the pump is dispensing properly by discharging about 1 µL of the collagenase before infusing into the animal; this also removes any additional air bubbles at the end of the syringe. By doing so, you ensure that the animal receives the entire desired amount of collagenase.

References

1. Mracsko E, Veltkamp R (2014) Neuroinflammation after intracerebral hemorrhage. Front Cell Neurosci 8:388

2. Qureshi AI, Mendelow AD, Hanley DF (2009) Intracerebral haemorrhage. Lancet 373:1632–1644

3. van Asch CJ et al (2010) Incidence, case fatality, and functional outcome of intracerebral haemorrhage over time, according to age, sex, and ethnic origin: a systematic review and meta-analysis. Lancet Neurol 9:167–176

4. Hemphill JC 3rd et al (2015) Guidelines for the management of spontaneous intracerebral hemorrhage: a guideline for healthcare professionals from the American Heart Association/American Stroke Association. In: Stroke

5. Kreitzer N, Adeoye O (2013) An update on surgical and medical management strategies for intracerebral hemorrhage. Semin Neurol 33:462–467

6. Xi G, Keep RF, Hoff JT (2006) Mechanisms of brain injury after intracerebral haemorrhage. Lancet Neurol 5:53–63

7. Keep RF et al (2005) The deleterious or beneficial effects of different agents in intracerebral hemorrhage: think big, think small, or is hematoma size important? Stroke 36:1594–1596

8. Davis SM et al (2006) Hematoma growth is a determinant of mortality and poor outcome after intracerebral hemorrhage. Neurology 66:1175–1181

9. Dowlatshahi D et al (2011) Defining hematoma expansion in intracerebral hemorrhage: relationship with patient outcomes. Neurology 76:1238–1244

10. Brouwers HB, Greenberg SM (2013) Hematoma expansion following acute intracerebral hemorrhage. Cerebrovasc Dis 35:195–201

11. Aronowski J, Zhao X (2011) Molecular pathophysiology of cerebral hemorrhage: secondary brain injury. Stroke 42(6):1781

12. Belur PK et al (2013) Emerging experimental therapies for intracerebral hemorrhage: targeting mechanisms of secondary brain injury. Neurosurg Focus 34:E9

13. Klebe D et al (2015) Modulating the immune response towards a neuroregenerative peri-injury milieu after cerebral hemorrhage. J Neuroimmune Pharmacol 10(4):576–586

14. Ma Q et al (2011) History of preclinical models of intracerebral hemorrhage. Acta Neurochir Suppl 111:3–8

15. Manaenko A et al (2011) Comparison of different preclinical models of intracerebral hemorrhage. Acta Neurochir Suppl 111:9–14

Chapter 8

A Rat Burn Injury Model for Studying Changes in Microvascular Permeability

Katie Wiggins-Dohlvik and Binu Tharakan

Abstract

The management of burn patients is an extremely complex and clinically challenging for patient care. Aside from the increasing reports of burn injury and morbidity and mortality directly related to it, the pathobiology of burn trauma is not clearly understood. The rat model of burn trauma described here is currently used in research laboratories to study various aspects of burn injury, including vascular dysfunctions. This model demonstrates the infliction of thermal injury in Sprague-Dawley rats using a well-established boiled water approach. We have utilized intravital microscopy to examine the microvascular hyperpermeability, the excessive leakage of proteins and fluids from the intravascular space to the extravascular space in mesenteric postcapillary venules using this model. An increase in microvascular permeability is a strong indicator of microvascular dysfunctions leading to tissue edema in burn trauma.

Key words Burn injury, Thermal injury, Burn trauma, Vascular hyperpermeability

1 Introduction

The management of burn trauma patients is an extremely complex process and is clinically challenging for patient care. Aside from the morbidity and mortality directly related to burn injury, the pathophysiology of burn trauma is equally deleterious [1]. For patients in the acute phase of burn injury, adequate fluid resuscitation is paramount to minimize further injury from such derangements; however, its importance is often equal to the difficulty of striking the appropriate balance between enough and too little fluid for resuscitation.

In such situations, delineated protocols outline resuscitation and include the Parkland, modified Parkland, Brooke, modified Brooks, Evan's, and Monafo's formulas [2]. Despite these guidelines and researches in the field, burn trauma patients often display clinical signs of underresuscitation such as lactic acidosis or acute kidney injury; over-resuscitation is equally problematic and can cause abdominal compartment syndrome or intense tissue swelling

Binu Tharakan (ed.), *Traumatic and Ischemic Injury: Methods and Protocols*, Methods in Molecular Biology, vol. 1717,
https://doi.org/10.1007/978-1-4939-7526-6_8, © Springer Science+Business Media, LLC 2018

[3–5]. The fluid leak and vascular dysfunction seen in burn injury is poorly understood but has tremendous clinical impact as it can render patients with both intense swelling and decreased intravascular volume. Better understanding of the process responsible for this may be highly beneficial in treating the innumerable patients afflicted with these derangements in civilian and military settings. The challenges of fluid management in burn patients stem from dysfunction on multiple fronts; however, dysregulation of the vascular system is thought to be paramount mainly due to increased vascular permeability and subsequent edema that commonly accompany severe thermal injuries [4, 6–11].

Investigation in vitro only examines isolated facets of the process. Clinical or in vivo experiments are not always feasible, intra vital microscopic investigation of vessels in real time offers unique and invaluable insight into the mechanisms of vascular leak seen in thermal injury. An increase in microvascular permeability, the excessive leakage of proteins and fluid from the intravascular space to the extravascular space, is a strong indicator of microvascular dysfunctions leading to tissue edema in burn trauma [8–11].

The use of a well-established and reproducible rodent model, as described in this protocol, is critical to fundamental research in this field. Several laboratories, including our laboratory, currently use this model to study various aspects of burn pathophysiology, with particular focus on the study of microvascular abnormalities and permeability changes associated with burn trauma [8–11]. While this is one of the most reliable approaches currently available for basic science researchers, further studies and model development including its automation may be critical to understand the cellular, molecular, and physiological changes associated with burn injuries and for therapeutic drug development against burn injury.

2 Materials

2.1 Animals

1. Sprague-Dawley rats (Charles River Laboratories Wilmington, MA) (*see* **Notes 1** and **2**).
   (a) Male.
   (b) 250–350 g.
   (c) Aged 5.5–9.0 weeks old.

2.2 Animal Preparation

1. Polyethylene tubing (PE-50, 0.58 mm internal diameter), two pieces approximately 20–30 cm in length, with one end beveled and the other attached to a three-way stop cock via a blunt tipped needle, flushed with injectable saline.
2. 5 cc syringe and 27 gauge ½" needle.
3. Urethane, 1.75 mg/kg, diluted in saline (*see* **Note 3**).

4. Four precut pieces 5-0 silk, 10–13 cm in length.

5. Bulldog clamp.

6. Electrocautery.

7. Heating pad.

2.3 Modified Walker/Mason Burn Model

1. Plastic mold.

2. Thermometer.

3. Timer.

4. Water.

5. Heating surface.

6. Container large enough to submerge dorsal aspect of animal.

7. Towel to dry animal.

2.4 Post Injury/Imaging

1. FITC albumin, (50 mg/kg); dissolved in injectable saline warmed to at least room temperature.

2. Sterile, injectable, sodium chloride 0.9% (normal saline, NS).

3. Electrocautery.

4. Petroleum ophthalmic ointment.

5. Cotton tipped applicators.

6. 2 × 2 squares of gauze.

7. Plexiglass mounting stage.

8. Plastic wrap, cut to approximately 10 × 10 cm.

9. Forceps.

10. Intravital microscope with 60× objective, camera, and display for viewing (we use a Nikon E 600 microscope, Tokyo, Japan; Nikon Instruments 60× water submersion objective, Inc., Natick, MA; Photometric Cascade Camera Roper Scientific, Tuscon, AZ respectively).

11. Software for image analysis (we use Nikon NIS Element Software).

12. Small centrifuge tubes.

3 Methods

Before starting, all experiments should be approved by the necessary institutional oversight committees in accordance with the relevant government regulations.

1. Animals should be housed in 12:12 h light–dark cycle with room temp 25 °C ± 2 °C and humidity constant 55% with free access to food and water until 12 h prior to experiments, after which animals should be fasted.

2. Animals should be anesthetized with intramuscular urethane (1.75 mg/kg, dose divided in two and administered in bilateral hindquarters).

3. After 30 min, anesthesia should be ensured with toe pinch.

4. Begin heating an appropriately sized container of water (so that 30% total body surface area can be completely submerged to induce burn).

5. Apply ophthalmic ointment to the animal's eyes to prevent dryness.

6. Shave the animal's neck and abdomen. Also shave the dorsum of the animal according to the desired total body surface to be burned.

7. A dime to quarter size area of skin overlying the lateral neck should be removed and the external jugular vein identified and isolated. 2 5-0 silk sutures should be placed around the vein but not tied.

8. Make a small venotomy with care taken to avoid complete transection.

9. Insert the beveled end of the polyethylene tubing carefully into the venotomy and secure with the 5-0 silk proximal and distal to the insertion site.

10. Next move medial and identify the carotid artery. Isolate the vessel with care and dissect away the vagus nerve. Place a small bulldog clamp as distal on the vessel as possible, taking care not to damage the vessel which can be very small and fragile. Next encircle the vessel with two of the precut 5-0 silk sutures but similar to the vein, do not tie them. Make a small arteriotomy and insert the beveled end of the polyethylene tubing. Secure with the silk sutures and remove the bulldog clamp.

11. Open the stopcock and ensure back flow from the carotid and flush with 0.5 cc saline to keep the line from clotting.

12. Carefully transfer the animal to the mold and secure so that the desired amount of skin is exposed on the dorsal aspect of the rat (*see* **Note 4**).

13. With a thermometer, ensure that the container of water is 100 °C for burn animals or 37 °C for sham animals.

14. Submerge the dorsal aspect of the animal in the water for 10 s.

15. Remove the animal from the water and pat-dry to stop continuation of burn.

16. Start a timer to ensure that the first images captured are at 30 min post burn.

17. Place the animal supine on a heating pad to ensure normothermia and secure upper and lower limbs with tape exposing the shaved abdomen.

18. Make a midline celiotomy and carefully enter the abdominal cavity, avoiding injury to underlying bowel or liver.

19. Place the animal in a lateral decubitus position on your plexiglass mounting stage, taking care not to dislodge your central venous catheter or arterial line.

20. Place 2–3 ml. of normal saline on your plexiglass stage next to the abdomen. There should be enough fluid to form a thin layer under the bowel without air bubble but not so much fluid that results in the animal lying in a puddle of fluid, as this can cause your animal to become hypothermic.

21. Use two cotton tipped applicators to eviscerate the bowel over your pool of saline and identify the distal ileum and ileocecal value (*see* **Note 5**).

22. Inspect the mesentery. Vessels suitable for monitoring may be visible as small hair like structures traversing the mesentery and in our experience vessels more on the interior of the mesentery (away from the edge of the bowel) are better for imaging. It is best to locate such a vessel/vessels and position them in the middle of the exteriorized loop.

23. Unfold a 2 × 2 piece of gauze and roll it along its length. Use this to anchor the bowel loop and position it for imaging.

24. Cover the bowel with several milliliters of normal saline to keep it moist.

25. Cover the bowel with plastic wrap and position such that there are no bubbles in the liquid above or below your mesentery. This will function essentially as your coverslip and also keep the bowel moist and stave off excoriation during your experiment.

26. Move your animal to your microscope, attach the arterial line to your monitor, and your central venous catheter to your IV fluids.

27. We use a submersion lens and apply water on top of the plastic wrap.

28. Locate a vessel approximately 20–35 μm wide for study: the same vessel will be used for the duration of the experiment (we found that isolated vessels with very little surrounding fat or lymphatics are ideal).

29. At 25–27 min from your time of burn, inject FITC-albumin through the central venous catheter over approximately 1–2 min and flush with 1 cc of NS.

30. At 30 min after burn injury, obtain your first intravital images.

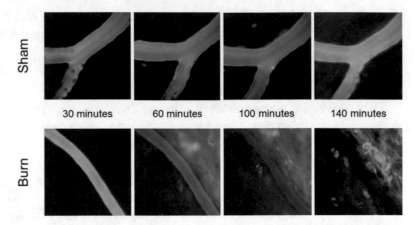

Fig. 1 Burn trauma-induced increase in vascular permeability in rat mesenteric postcapillary venules, studied using intravital microscopy. FITC-albumin was used as a fluorescent tag. The changes in vascular permeability was determined based on the changes in fluorescent intensity measured and calculated intra and extra-vascularly (data not included). Images representing a sham and burn animal starting from 30–140 min is shown. Increased FITC-albumin extravasation indicating microvascular hyperpermeability is observed in burn rats compared to sham rats

31. Utilizing the same vessel, obtain images per your predetermined protocol (we obtain images at 30, 60, 80, 100, 120, and 180 min after burn injury for the majority of our experiments) (Fig. 1).

32. Resuscitate the animal during the experiments as needed (*see* **Notes 6** and **7**).

33. At the conclusion of your experiment, animals are euthanized by exsanguination via arterial line (*see* **Note 8**).

34. Images are analyzed with Nikon NIS Element Software, differences from baseline.

4   Notes

1. Younger male rats weighing 250–350 g were utilized for our experiments as heavier, older animals have increased mesenteric fat, making visualization of small vessels with intravital microscopy more difficult.

2. Animals were given free access to food and water and then fasted for 12 ours before experiments. We found this to be helpful for two reasons. First, there was less peristalsis within the bowel and this aided in obtaining clear images. Additionally, a lower stool burden allows more simple exteriorization of bowel and easier coverage with plastic film, which aids in keeping the bowel moist for the duration of experiments.

3. Urethane was used for anesthesia and analgesia because, compared to more commonly used agents, it has very limited effect on cardiovascular status or vessel permeability, making it ideal for experiments examining these parameters. It can be carcinogenic and therefore should be handled with care and mixed under a fume hood.

4. Take care during mobilizing animals to ensure that central venous access and arterial lines do not become dislodged.

5. Touching the mesentery directly can cause injury to vessels and vasospasm and should be avoided throughout experiments.

6. IV fluid resuscitation vary with experiments. In several of our experiments we conducted fluid guided resuscitation and only gave animals fluids when their mean arterial pressure fell below a preset threshold [3–5]. During other experiments animals were give a preset amount of IV fluid per kilogram per hour [2]. The goals of your experiment should dictate which method is utilized. Nonetheless, the central venous catheter should be utilized in some manner intermittently for fluid infusion or it will clot.

7. For the duration of the experiment, ensure that the bowel is covered with plastic wrap and kept moist to avoid excoriation as this can lead to vessel reactivity and interference with examination of such vessels.

8. In our laboratory, shed blood collected from exsanguination of animals at the conclusion of experiments is commonly centrifuged at $4266 \times g$ for 20 min, and then the serum removed and frozen at $-80\ °C$. Tissue is also harvested, flash-frozen in liquid nitrogen, and stored at $-80\ °C$ for use in additional experiments.

References

1. Alvarado R, Chung KK, Cancio LC, Wolf SE (2009) Burn resuscitation. Burns 35:4–14

2. Haberal M, Abali AE, Karakayali H (2010) Fluid management in major burn injuries. Indian J Plast Surg 43(Suppl):S29–S36

3. Dries DJ (2009) Management of burn injuries – recent developments in resuscitation, infection control and outcomes research. Scand J Trauma Resusc Emerg Med 17(14):1–13

4. Greenhalgh DG (2007) Burn resuscitation. J Burn Care Res 28:555–565

5. Pham TN, Cancio LC, Gibran NS (2008) American Burn Association practice guidelines burn shock resuscitation. J Burn Care Res 29:257–266

6. Demling RH (2005) The burn Edema process: current concepts. J Burn Care Rehabil 26:207–227

7. Walker H, Mason A (1968) A standard animal burn. J Trauma 8:1049–1051

8. Stagg HW, Whaley JG, Tharakan B, Hunter FA, Jupiter D, Little DC, Davis ML, Smythe WR, Childs EW (2013) Doxycycline attenuates burn-induced microvascular hyperpermeability. J Trauma Acute Care Surg 75:1040–1046

9. Wiggins-Dohlvik K, Oakley RP, Han MS, Stagg HW, Alluri H, Shaji CA, Davis ML, Tharakan B (2016) Tissue inhibitor of metalloproteinase-2 inhibits burn-induced derangements and hyperpermeability in microvascular endothelial cells. Am J Surg 211:197–205

10. Wiggins-Dohlvik K, Han MS, Stagg HW, Alluri H, Shaji CA, Oakley RP, Davis ML, Tharakan B (2014) Melatonin inhibits thermal injury-induced hyperpermeability in microvascular endothelial cells. J Trauma Acute Care Surg 77:899–905

11. Wiggins-Dohlvik K, Stagg HW, Han MS, Alluri H, Oakley RP, Anasooya Shaji C, Davis ML, Tharakan B (2016) Doxycycline attenuates lipopolysaccharide-induced microvascular endothelial cell derangements. Shock 45:626–633

Chapter 9

Modeling Transient Focal Ischemic Stroke in Rodents by Intraluminal Filament Method of Middle Cerebral Artery Occlusion

Mary Susan Lopez and Raghu Vemuganti

Abstract

The middle cerebral artery occlusion (MCAO) model is widely used for inducing a focal cerebral ischemic insult (stroke) in rodents. Here, we describe a method for transient MCAO technique without craniotomy in both mice and rats. In our laboratory, this technique yields consistent secondary brain damage that evolves over a period of 3–7 days of reperfusion after transient MCAO. We also describe the methods for analyzing postischemic motor dysfunction and infarct volume in rodents subjected to transient MCAO.

Key words Middle cerebral artery occlusion, Focal cerebral ischemia, Transcardiac perfusion, Beam-walk test, Rotarod test, Sticker removal test, Infarct volume

1 Introduction

The ability to induce a controlled and consistent ischemic insult in experimental animals allows studying the cellular and molecular mechanisms of poststroke brain damage and testing of new therapeutic drugs. Middle cerebral artery occlusion (MCAO) has been a valuable model of focal ischemic stroke since 1981, when Tamura et al. [1] demonstrated permanent occlusion in the distal MCA via craniectomy. This method and its variations are still widely used in rodents, but does not allow for analysis of reperfusion-mediated brain damage. A transient MCAO model was first developed in rats by Koizumi et al. in 1986 [2], and was later expounded upon by Longa et al. [3]. The Koizumi method uses a silicone-tipped nylon monofilament, which is inserted into the common carotid artery (CCA) and advanced through the internal carotid artery (ICA) to the origin of the MCA. The Longa method employs a similar intraluminal filament technique, but the filament is inserted into the external carotid artery (ECA) instead of the CCA. In both cases, reperfusion occurs upon withdrawing the filament, but in

Binu Tharakan (ed.), *Traumatic and Ischemic Injury: Methods and Protocols*, Methods in Molecular Biology, vol. 1717, https://doi.org/10.1007/978-1-4939-7526-6_9, © Springer Science+Business Media, LLC 2018

the Koizumi method the CCA will be permanently ligated while the Longa method allows for its reopening. Thus, the Koizumi method can be employed only in rodents with well-developed circle of Willis that permits adequate blood flow in the MCA, whereas the Longa method can be employed in any rodent [4].

The following are the methodological details of the intra-luminal filament model of MCAO in mice (Koizumi method) and rats (Longa method), including postprocedural analysis of behavior and infarct volume as developed and routinely used in our laboratory.

2   Materials

The surgical tools described below are adequate to perform this procedure, but the surgeon may opt to add or substitute certain tools as desired.

2.1 Anesthesia Rig and Surgical Area

1. Surgical equipment (Fig. 1a).
2. Oxygen cylinder with gas limiter (Airgas).
3. Nitrous oxide cylinder with gas limiter (Airgas).
4. Isoflurane.
5. Neck rest. A roll of gauze taped into a cylinder can be used as an appropriate neck rest.
6. Heated recovery cage for occlusion and reperfusion.
7. Electric razor.

2.2 Supplies for Sterile Technique

1. Bead sterilizer (Fine Science Tools, #18000-45).
2. Sterile pads or stockings.
3. Iodine.
4. Sterile scrub brushes.
5. Sterile towels.
6. Sterile gloves.
7. Surgical mask.
8. Surgical bonnet.
9. Surgical gown.

2.3 Surgical Tools

1. Surgical tool set (Fig. 1b, #1-19).
2. Microscissors (Fine Science Tools, #91500-09 and #15000-00). The larger pair (Fig. 1a) is used for cutting through fascia and connective tissue and the smaller pair (Fig. 1a) is used for arteriotomy. This keeps the arteriotomy scissors sharp, and the

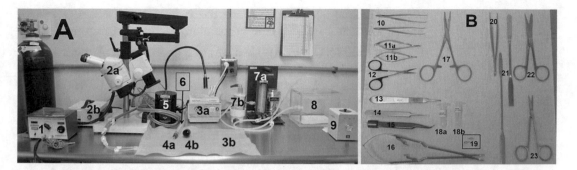

Fig. 1 Surgical area and equipment for MCAO. (**a**) Surgical table with anesthesia rig. Equipment pictured: (*1*) Electric cauterizer (Radionics, Model 440E bipolar coagulator). (*2a*) Surgical microscope (Leica MZ6). (*2b*) Light box for surgical microscope. (*3a*) Warm water blanket (Stryker TP22C). (*3b*) T-pump system (Gaymar T/Pump T-500). (*4a*) Mouse nose cone (Vetequip). (*4b*) Rat nose cone (Vetequip). (*5*) Prehensile light source (Dolan Jenner, Fiber-lite 190). (*6*) Bain circuit (Surgivet). (*7a*) Isoflurane vaporizer (Surgivet). (*7b*) Isoflurane charcoal filter (Vetequip). (*8*) Induction chamber (Surgivet). (*9*) Bead sterilizer (Fine Science Tools, #18000-45). Additional equipment is described in Subheading 2.1. (**b**) Surgical tools for MCAO and transcardiac perfusion. Tools pictured: (*10*) Pair of curved, extra fine Graefe forceps (Fine Science Tools, #11152-10). (*11a*) Vannas large microscissors for sharp dissection (Fine Science Tools, #91500-09). (*11b*) Vannas small microscissors for arteriotomy (Fine Science tools, #15000-00). (*12*) Iris scissors (Fine Science Tools, #14058-09). (*13*) Scalpel holder (blades not pictured) (Fine Science Tools, #10003-12). (*14*) Straight, extra fine, sharp-tip forceps (Fine Science Tools, #11295-10). (*15*) Cauterizing forceps (Radionics). (*16*) Microsurgical clip applicator (Mizuho, #07-942-02). (*17*) Needle driver (Fine Science Tools, #12002-14). (*18a*) MCAO filaments for mice (Doccol, #6023). (*18b*) MCAO filaments for rats (Doccol, #403934). (*19*) Microsurgical clips (Mizuho, #07-940-56). (*20*) Rat-toothed forceps (Fine Science Tools, #91127-12). (*21*) Spatulae. (*22*) Sharp/blunt surgical scissors for rat perfusion (Fine Science Tools, #14001-13). (*23*) Fine, pointed surgical scissors for mouse perfusion (Fine Science Tools, #14068-12). Additional detail on certain tools can be found in Subheadings 2.3 and 2.5. Some tools not pictured are described in Subheadings 2.3 and 2.5

size affords greater precision. However, the surgery is possible with only the larger pair of microscissors.

3. Microsurgical clip applicator and clips (Mizuho, #07-942-02; clips #07-940-56). This size can be used only in rats. Microvascular clips (Fig. 1b) tend to limit visibility, but can be useful during emergencies. For example, if the ICA ligature comes loose after arteriotomy, a clip can be placed near the arteriotomy so the ligature can be retied. Some surgeons use clips in lieu of thread ligatures in rats, but this is not recommend as it reduces the visibility.

4. Retractors (surgeon's choice). Retractors enhance visibility while conducting the surgery in rats, but difficult to use in mice. An alternate to using retractors is to loop pieces of silk suture around the arteries to demarcate where they are under the glandular tissue. If this approach is taken, it is advisable to have hemostats available to keep the silk organized. Another alternate approach is to tie back the digastric and/or sternomastoid muscles directly, using microsurgical needles with

attached silk. But, this approach is not advised as it can be traumatic and difficult to maintain aseptic technique.

5. Filaments (Doccol). Choosing an appropriate filament is critical for proper occlusion of MCA. Premade filaments (rats: Doccol 4-0 #403934 and mice: Doccol 6-0 #6023) are observed in our laboratory to be consistent and effective. If desired, homemade filaments can be used, but some trial and error is involved. For some guidance, *see* [5, 6] or the Doccol website (www.Doccol.com).

6. Cotton-tipped applicators (Puritan, 806-WC) are useful for blunt dissection, stemming accidental bleeding, moistening regions with sterile saline, and applying topical gel analgesics. This tool should not be omitted.

7. 6-0 braided nonabsorbable silk thread (Fine Science Tools, #18020-60). Sterile 6-0 nonabsorbable silk thread is difficult to find. For large spools of thread, cut into lengths and follow manufacturer's instructions for sterilization.

8. Curved needles with attached braided nonabsorbable silk sutures (Ethicon; surgeon's choice).

9. 1 mL syringes with needles.

10. Analgesics: Bupivacaine and lidocaine.

11. Sterile Saline: 0.9% NaCl solution.

2.4 Equipment for Assessing Post-MCAO Behavior

1. Beam-walk apparatus: This is usually homemade with sealed wood. As long as the animal travels 1 m during the test, and the same apparatus is used for all animals, the behavior data will be internally consistent. Some trial and error may be required for initially building this apparatus.

2. Rotarod (Letica Scientific Instruments, Rotarod R/S LE8500).

3. Adhesive stickers: 1 cm × 1 cm for rat, 0.3 cm × 0.3 cm for mouse (Avery).

2.5 Tools for Transcardiac Perfusion, TTC Staining, and Long-Term Brain Storage

1. Perfusion tool set (Fig. 1b, #20-23).

2. Perfusion syringe: 60 mL with 22 G needle for rat, 10 mL with 26 G needle for mouse.

3. Chilled Saline: 0.9% NaCl solution.

4. Tub of crushed ice.

5. Brain matrix (ASI Instruments, rat: #RBM-4000C, mouse: #RBM-2000C).

6. 2,3,5-Triphenyltetrazolium chloride (TTC, Sigma, T8877). Keep TTC solution in dark. A 0.25% TTC solution provides excellent staining if sections are incubated for 30 min. *See* [7] for additional guidance if desired.

7. 6 or 12 well tissue culture plates (not sterile).

8. 37 °C incubator. If keeping an incubator in the animal sacrifice area is impractical or infeasible, wrap the plate in the warm water mat set to 37 °C.

9. 4% paraformaldehyde solution.

10. 30% sucrose solution made in PBS.

11. Scintillation vials.

12. Perfusion apparatus (e.g., Leica Perfusion One or AutoMate Rodent Fixation System).

3   Methods

Adhere to aseptic technique for the entirety of the surgery. Prior to surgery, sterilize tools using ethylene oxide followed by autoclaving or bead sterilization. Follow institutional guidelines for laboratory animal surgery and care.

3.1 Preparation and Anesthesia

1. Adjust the oxygen level on the flowmeter to 1 L/min and nitrous oxide to 0.3 L/min. Adjust the isoflurane level to 5% on the vaporizer. These flowmeter numbers are guidelines that apply to most vaporizer units.

2. Place the animal in the induction chamber until the surgical plane is reached (2–5 min), then adjust the isoflurane level to 2%. Shave rats before surgery in an area separate from the surgery area and clear any loose fine hair after the shaving. For mice, it is preferred to sterilize the fur rather than shaving [4] as the hair is fine and difficult to clear after shaving.

3. Adjust the animal in a supine position on a neck rest so that the neck is elevated, but the trachea is unbent. Place a thin piece of material under the Bain circuit to facilitate breathing, if necessary. Ensure that the surgeon has clear access to the surgical area at all times and adjust the body if needed.

4. Sterilize the surgical region three times with iodine, starting in the center near the intended incision, and moving outward in a circular motion. Cover the animal with a sterile drape.

5. Scrub up to the elbow with hot water and an iodine scrub brush. Dry off with sterile towels, then don sterile gloves, gown, mask, and bonnet.

3.2 Exposing the Carotid Arteries

1. Inject bupivacaine (0.1 mL) subcutaneously along the intended incision, then make a midline incision in the ventral side of the neck. The incision can be made with a scalpel (rats) or scissors (mice). Take care to avoid crushing the trachea or cutting through interior tissue.

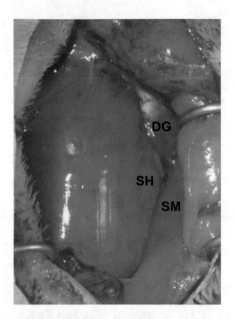

Fig. 2 Muscular intersection overlaying the carotid arteries. This subtle triangle formed by the sternohyoid (SH), digastric (DG), and sternomastoid (SM) muscles is pulled apart to reveal the carotid arteries underneath

2. Under dissecting microscope, open the skin, push aside glandular and connective tissue and use microscissors to cut the layer of connective tissue down the midline, between the large mandibular salivary glands. Push the mandibular salivary glands laterally with a sterile cotton-tip swab to expose the triangular intersection of sternohyoid, sternomastoid, and digastric muscles (Fig. 2). This is not necessary in mice, as the triangular intersection will be visible after the incision.

3. Using microscissors, gently dissect the fascia connecting the muscles, then retract them using blunt dissection (this will take some force; spinning the swabs away from the midline will help pull the tissues apart). This will reveal the CCA, ICA, and ECA (Fig. 3a).

4. Free the CCA, ICA, and ECA from the underlying tissue using the Graefe forceps. Take care not to damage the vagus nerve. The carotid arteries are bound by web-like connective tissue on all sides, but the bifurcation area is bound by a thicker connective tissue. Free the CCA first, by inserting closed Graefe forceps underneath one side of the artery and moving back and forth to break the web-like connective tissue. Isolate the CCA from the vagus nerve gently by inserting closed Graefe forceps underneath the CCA, and gently poking the end of the forceps between the CCA and vagus nerve on the opposite side. Open and close the forceps until a small opening is made, then

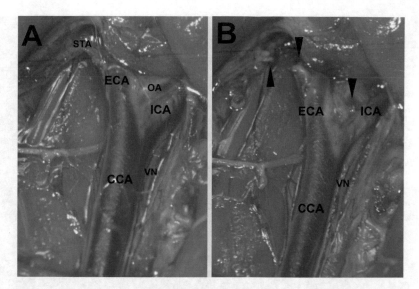

Fig. 3 Anatomy of rat carotid arteries. (**a**) Visualization of common carotid artery (CCA), external carotid artery (ECA), internal carotid artery (ICA), superior thyroid artery (STA), vagus nerve (VN), and occipital artery (OA) in rat. (**b**) Carotid artery region after OA and STA cauterization. *Arrowheads* in (**b**) indicate cauterized stumps

employ the back-and forth motion to gently separate the two. Isolate the ICA and CCA in a similar fashion.

5. Rat only: cauterize and cut the occipital artery (OA) and the superior thyroid artery (STA). Isolate the ECA as far rostral as possible (the hyoid bone will limit the dissection). Isolate the STA and OA (Fig. 3a) and cauterize them completely (Fig. 3b).

3.3 Introducing the Filament

1. Ligate the ECA using 6-0 silk thread. Rat: ligate the ECA as rostrally as possible. Mouse: ligate the ECA near the bifurcation of the CCA.

2. Using microvascular clips or 6-0 silk thread, ligate the CCA and ICA. In order to have sufficient working space, ligate the CCA as caudally as possible and ligate the ICA as rostrally as possible. Rat: transiently ligate both arteries. Mouse: permanently ligate the CCA (take care not to twist CCA) and transiently ligate the ICA.

3. Use 6-0 silk thread to loosely tie a transient ligature around the filament insertion site. Rat: tie around the ECA. Mouse: tie around the CCA, between the permanent ligature and the CCA bifurcation.

4. Make a partial arteriotomy using microscissors. Rat: cut near the permanent ligature on the ECA (Fig. 4a). Mouse: cut near the permanent ligature on the CCA. Do not lift the vessel to perform the arteriotomy, as this can flatten or twist the vessel.

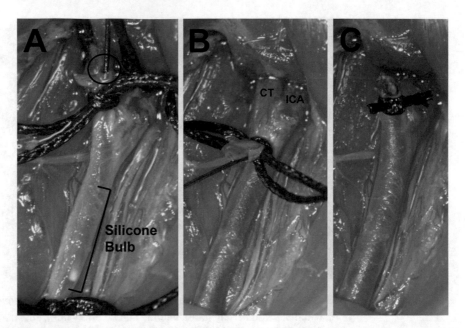

Fig. 4 Inserting and withdrawing the filament using Longa's method. (**a**) Filament just after insertion into the ECA. The arteriotomy region is *circled*. The silicone bulb is visible from outside the artery. (**b**) Occlusion of MCA. The connective tissue (CT) near the bifurcation of the common carotid is intact in the figure, but the surgeon may wish to cut this tissue away. (**c**) Reperfusion. The ECA has been cauterized and permanently ligated. Blood flow from CCA is restored

5. Insert the filament into the arteriotomy and tighten the transient ligature from **step 3** around the filament (Fig. 4a). Keep the ligature tight enough to prevent bleeding, but loose enough to allow the filament to advance. Inserting the filament is a difficult step and we recommend using the straight-tip pointed forceps as a guide to open the vessel while inserting the filament.

6. Advance the filament up to the ICA ligature/clip. Rat only: cut ECA at arteriotomy site to free the ECA stump, then align the stump with the ICA and advance the silicone tip up to the ICA ligature. We recommend keeping the silicone bulb inside the ECA trunk, then aligning the bulb so that it is parallel with the CCA. Angle the bulb above the ICA so that the filament can be gently pushed rostrally into the ICA. An alternate method is to advance the bulb to the CCA, then flip the bulb inside the vessel so that it points into the ICA. This technique is more difficult when using filaments with longer silicone bulbs.

7. Slowly open the ICA ligature and advance the filament through the ICA. Watch the silicone tip inside the vessel as it advances, and make sure that the bulb does not travel down the pterygopalatine artery (PPA). In the event that the filament travels down the wrong artery, simply withdraw back into the ICA and

start again. Look for the PPA branch off of the ICA, and physically direct the bulb using Graefe forceps if necessary. You can ligate the PPA to prevent the filament from entering this artery.

8. Once the filament has reached the MCA region, slight resistance will be felt. Tighten the ligature around the filament and record the time as the start of ischemia (Fig. 4b). The origin of MCA lies 18–20 mm in rats and 9–11 mm in mice from the CCA bifurcation. In both cases, if using Doccol filaments, the bulb will not be visible when occlusion is achieved. Ensure that the filament ligature is tied tightly in place as it may get dislodged when the animal becomes conscious and active during the ischemic period. Rat only: remove the transient ligature from the CCA.

9. Moisten the area with sterile saline, apply lidocaine as a topical analgesic and close the wound with sutures or staples. Apply bupivacaine along the sutures after wound closure and place the animal in a temperature controlled recovery chamber. Monitor for abnormal behavior other than that due to ischemic insult.

3.4 Reperfusion and Postsurgical Care

1. Reanesthetize and reposition the animal for surgery and resterilize the area as described in Subheading 3.1.

2. Reopen the neck incision to reveal the filament. Gently pull the filament out of place until the bulb is visible, and record the time as the end of ischemia.

3. Bring the bulb down to the arteriotomy and transiently religate the ICA. Rat only: transiently religate the CCA as well.

4. Loosen the transient ligature around the filament and gently remove it, then permanently tighten the ligature around the arteriotomy to prevent bleeding. In rats, permanent ligature can be reinforced by cauterizing the ECA trunk (Fig. 4c). In mice, permanent ligature can be reinforced by tightening the ICA ligature.

5. Rat only: Remove transient ligature from the ICA and CCA (Fig. 4c).

6. Moisten the area with sterile saline, apply lidocaine as a topical analgesic and close the wound with sutures or staples. Inject isotonic saline intraperitoneally (17 mL/Kg) to prevent dehydration. Apply bupivacaine along the sutures after wound closure and place the animal in a temperature controlled recovery chamber. Monitor its behavior for 1–2 h.

7. Return the animal to a housing cage. Place moistened rodent chow on the bottom of a cage to facilitate eating. Ischemic animals should not be housed with nonischemic animals as they may harm the ischemic rodents.

8. Monitor the animal on the day after surgery. Look for signs of hemiplegy that include impaired grooming, stasis and circling. Impaired grooming and dehydration may result in urethral blockages in mice which usually fall out or get dissolved. But in extreme cases they need to be physically removed and the animal should be rehydrated with an injection of sterile saline.

3.5 Postischemic Functional Testing

When planning to use a functional test, ensure that the animals have been trained daily for at least 3 days prior to the induction of ischemia. To avoid bias, use a blinded evaluator for scoring these tests.

1. Neurological deficit scoring: We use a 5-point scale for measuring neurological deficits after MCAO [3, 8]. The scale will be "0" for no deficits (normal behavior), "1" for mild deficit (failure to fully extend contralateral forepaw), "2" for moderate deficit (contralateral circling), "3" for severe deficit (contralateral falling or flexion), and "4" for critical deficit (no spontaneous movement and depressed consciousness).

2. Beam-walk test: Trained animals can walk across a narrow beam for 1 m [8]. Ischemic animals will have difficulty performing this task as the contralateral rear paws slide off the beam indicating foot faults which can be counted to assay motor function. The time to complete the task can also be recorded to assay anxiety and motor function.

3. Rotarod test: Trained animals can stay on a slowly rotating rod (4–8 rpm) for a given period of time (3–5 min) [8]. Ischemic animals will have difficulty performing this task and will fall off of the rotating rod. The length of time the animal can stay on the rotating rod can be recorded to assay motor function.

4. Adhesive sticker removal test: When small circular adhesive stickers are placed on forepaws, normal trained rodents can quickly sense and remove them [8, 9]. Ischemic animals demonstrate difficulty in sensing and removing the stickers. The length of time taken to sense and to remove the sticker can be recorded to assay sensory-motor function.

3.6 Transcardiac Perfusion

1. Induce deep anesthesia with isoflurane and place the animal on ice. Be certain the animal is fully unresponsive using pedal and tail pinch tests. This is a terminal procedure, so overexposure to isoflurane is acceptable so long as the heart continues to beat throughout the procedure.

2. Use rat-toothed forceps to grab the abdomen at the base of the rib cage. Tent the tissue and cut laterally through the abdominal muscle. Once an abdominal incision is made, the procedure must be finished within 5 min. Be sure to cut all the way across the body to permit maximum blood drainage.

3. Grasp the xiphoid process with rat-toothed forceps and cut through the diaphragm along the rib cage. Cut rostrally through the lateral-most aspects of the rib cage, pulling the xiphoid process toward the head until the heart is exposed and the ribcage can be reflected up over the animal's neck. Use a hemostat to keep the rib cage open, if desired.

4. Insert a syringe (60 mL syringe for rat; 10 mL syringe for mouse) filled with chilled normal saline into the left ventricle and slowly administer the saline. Using scissors or the rat-tooth forceps, tear or cut a hole in the right auricle.

5. Continue to perfuse the animal until signs of blood loss are apparent: the lungs and paws will turn white, the auricle will pump out saline, and the liver will turn from dark red to pale tan. This should take no more than 2–3 min, and may not require all of the saline in the syringe.

6. After satisfactory blood loss, decapitate the animal and set the head on a sturdy surface for brain removal. At this point, shut off the anesthesia.

7. Using the scalpel blade, score the skull along the sagittal suture, moving all the way from the rostral-most aspect to the caudal-most aspect of the skull.

8. Using scissors, cut any remaining neck tissue off of the skull, and cut along the rear midline to make a single open line from the foramen magnum, past lambda and bregma, up to the snout.

9. Insert the closed point of the scissors into the snout near the nasal turbinates. Open the scissors slightly, and twist to open the skull along the scores that were made earlier. Open the scissors the rest of the way to force the skull open.

10. Invert the skull and with a chilled spatula gently dislodge the brain from the cranium. Invert the skull so that gravity helps to remove the brain. It is necessary to disrupt the cranial nerves and remove the brainstem from the foramen magnum. Following transient MCAO, infarction is largely seen in the cerebral cortex and striatum; hence take care not to touch the area between lambda and bregma with the spatula. Use the cerebellum and olfactory bulbs as leverage to remove the brain from the cranium, if necessary.

3.7 Infarct Volume Estimation by TTC Staining

1. Chill the rodent brain matrix on ice. Use the olfactory bulbs and cerebellum to position the brain inside the matrix. If the brain is fresh (as described), insert razor blades every 2 mm, then insert more razor blades between the first set to obtain 1 mm slices. This maintains the architecture and prevents "squishing" the slices. Some labs prefer freezing the brain at $-20\ ^\circ C$ for 2–3 min before placing it inside the matrix. If the

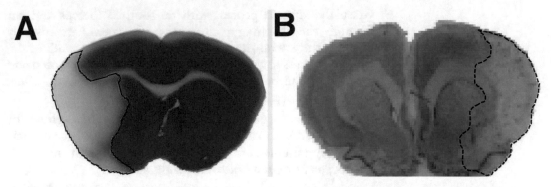

Fig. 5 Means of visualizing infarctions after MCAO. (**a**) Coronal brain section (1 mm thick) stained with TTC from representative adult mouse subjected to 60 min MCAO and 3 days of reperfusion. *Solid black line* surrounds the infarct region. (**b**) Coronal brain section (40 μm thick) stained with Cresyl Violet from representative adult rat subjected to 60 min MCAO and 7 days of reperfusion. *Dotted black line* surrounds the infarct region

brain is slightly frozen, then razor blades can be placed every 1 mm without fear of disrupting the architecture.

2. Gently remove the brain slices and place them into prefilled wells of a 6 or 12 well plate containing TTC solution. Incubate for 30 min at 37 °C.

3. After 30 min, remove slices from the TTC and place them in 4% PFA for storage (Fig. 5a).

4. Scan brain slices, find infarct area of each slice using ImageJ (http://imagej.nih.gov/ij) or equivalent program, and estimate infarct volume [10, 11].

5. An approximate volume can be estimated using rectangular summation ($V = d\left[\sum_{i=1}^{n-1} y_i\right]$, where d is the distance between sections, y is the area of section i and n is the total number of sections), such that the distance between the sections being analyzed is multiplied by the area of each section (typically, area × section thickness). Cavalieri's estimator [11] is an improvement on rectangular summation as it corrects for overprojection when section thickness is not negligible ($V = d\left[\sum_{i=1}^{n} y_i - (t)y_{max}\right]$, where t is the thickness of the section with the maximal area (y_{max})).

3.8 Long-Term Storage for Immunohistochemistry or Cresyl Violet (CV) Staining

1. Prepare the perfusion apparatus. Fill the apparatus with 4% PFA and ensure that there are no bubbles in the tubing.

2. Perform transcardiac perfusion as described in **steps 1–5** of Subheading 3.6.

3. Following saline perfusion, exchange the saline syringe for the perfusion apparatus and administer 4% PFA solution into the left ventricle. At this point, the anesthesia can be shut off.

4. Continue to perfuse until animal begins to exhibit tetany-like seizures and the limbs become difficult to move. It should be completely stiff by the time it is fully fixed.

5. Remove the skull using the method described in **steps 7–10** of Subheading 3.6. The skull will be considerably more difficult to break through after fixation, so rongeurs can be employed. As the brain is fixed, time is not of the essence and the surgeon can focus on preserving the brain's architecture.

6. Place the brain in a scintillation vial filled with 4% PFA, and keep at 4 °C overnight.

7. After the overnight fixation, transfer the brain to another scintillation vial filled with 30% sucrose and keep at 4 °C overnight.

8. Replace the sucrose every day until the brain sinks (should take 3–5 days), indicating that it has become fully dehydrated. At this point, transfer the brain to a scintillation vial filled with fresh 30% sucrose for long-term storage at 4 °C. The fixed, dehydrated brain can be cut with a microtome for immunohistochemistry and infarction analysis by Cresyl Violet staining (Fig. 5b).

References

1. Tamura A, Graham DI, McCulloch J, Teasdale GM (1981) Focal cerebral ischaemia in the rat: 1. Description of technique and early neuropathological consequences following middle cerebral artery occlusion. J Cereb Blood Flow Metab 1:53–60

2. Koizumi J, Yoshida Y, Nakazawa T, Ooneda G (1986) Experimental studies of ischemic brain edema 1. A new experimental model of cerebral embolism in rats in which recirculation can be introduced in the ischemic area. Jpn J Stroke 8:1–8

3. Longa EZ, Weinstein PR, Carlson S, Cummins R (1989) Reversible middle cerebral artery occlusion without craniectomy in rats. Stroke 20:84–91

4. Rousselet E, Kriz J, Seidah NG (2012) Mouse model of intraluminal MCAO: cerebral infarct evaluation by Cresyl violet staining. J Vis Exp 69

5. Tureyen K, Vemuganti R, Sailor KA, Dempsey RJ (2005) Ideal suture diameter is critical for consistent middle cerebral artery occlusion in mice. Neurosurgery 56:196–200

6. Belayev L, Alonso OF, Busto R, Zhao W, Ginsberg MD (1996) Middle cerebral artery occlusion in the rat by intraluminal suture. Neurological and pathological evaluation of an improved model. Stroke 9:1616–1622

7. Joshi CN, Jain SK, Murthy PSR (2004) An optimized triphenyltetrazolium chloride method for identification of cerebral infarcts. Brain Res Protocol 13:11–17

8. Nakka VP, Lang BT, Lenschow DJ, Zhang DE, Dempsey RJ, Vemuganti R (2011) Increased cerebral IGSylation after focal ischemia is neuroprotective. J Cereb Blood Flow Metab 31:2375–2384

9. Bouet V, Boulouard M, Toutain J, Divoux D, Bernaudin M, Schumann-Bard P, Freret T (2009) The adhesive removal test: a sensitive method to assess sensorimotor deficits in mice. Nat Protoc 4:1560–1564

10. Rosen GD, Harry JD (1990) Brain volume estimation from serial section measurements: a comparison of methodologies. J Neurosci Methods 35:115–124

11. Uylings HBM, van Eden CG, Hofman MA (1986) Morphometry of size/volume variables and comparison of their bivariate relations in the nervous system under different conditions. J Neurosci Methods 18:19–37

A Complete Guide to Using the Endothelin-1 Model of Stroke in Conscious Rats for Acute and Long-Term Recovery Studies

Hima C.S. Abeysinghe and Carli L. Roulston

Abstract

Multiple methods exist to model permanent and transient ischemia under anesthesia in animals, however most human strokes occur while conscious. The use of endothelin-1 as a vasoconstrictor applied to the perivascular surface of the middle cerebral artery is one of the only methods for inducing stroke in conscious animals. Here, we describe standard operating procedures for stereotaxic placement of an ET-1 guide probe above the middle cerebral artery, induction of stroke in conscious rats, predictive outcome scoring during stroke, and neurological behavioral tests that we use to monitor transient and continuing deficits. The inclusion of long term neurological assessment is of particular importance when taking into consideration the effects of stroke on brain remodeling.

Key words Standard operating procedure, Stereotaxic surgery, Conscious stroke, Predictive outcome, Neurological assessments

1 Introduction

Given the setbacks associated with neuroprotective drugs in clinical practice, it is clearly important to use animal models of stroke that mimic more closely the human condition. Endothelin-1 (ET-1) is a potent and long-acting venous and arterial constrictor. Originally isolated [1] and generated by endothelial cells, ET-1 exerts its effects through two main receptors, endothelin-A receptor (ETA-R) and endothelin-B receptor (ETB-R) [2, 3]. Targeting the receptor located on the perivascular surface of the middle cerebral artery lead to the development of the first conscious model of stroke in animals [3]; stereotaxic injection of ET-1 adjacent to the middle cerebral artery (MCA) in conscious rats resulted in constriction of the MCA followed by gradual reperfusion. This was the first viable model of conscious focal cerebral ischemia that resulted in cerebral infarctions equivalent to those observed using anesthetized models of MCAo *(3 & 13)*. Since then

Binu Tharakan (ed.), *Traumatic and Ischemic Injury: Methods and Protocols*, Methods in Molecular Biology, vol. 1717,
https://doi.org/10.1007/978-1-4939-7526-6_10, © Springer Science+Business Media, LLC 2018

application of ET-1 to the MCA to induce focal ischemic strokes has also been adapted for use in primates [4], representing a most relevant platform for clinical translation to investigate ischemic stroke.

ET-1 induced reduction in CBF with gradual reperfusion resembles the majority of human strokes where varying degrees of spontaneous reperfusion occurs in the absence of thrombolysis [5]. The model enables prediction of outcomes based on observed behavioral changes as they occur during stroke induction. Similar prognostic clinical approaches, such as the use of the Scandinavian Stroke Scale, allow prediction of functional outcome and survival of stroke sufferers in order to correctly stratify treatment groups in clinical trials [6]. This same approach has now been developed for use in rats [7] where concurrent observations during ET-1 stroke induction can be used to accurately predict the extent of histological damage and neurological deficits incurred [8]. Such a predictive outcome model enables stratification of rats into treatment groups ensuring that stroke severity is evenly represented across all treatments to be assessed [7, 8]. This holds particular relevance to the clinical setting since humans are not preassigned to treatment groups prior to the onset of stroke, and not all human stroke is the same [6]. Historically, preclinical animal studies randomized rats into treatment groups prior to stroke induction which may have led to groups being unevenly weighted according to stroke outcome.

A high degree of variability in stroke damage is observed between individuals and this too can be accounted for when using the ET-1 model since all degrees of stroke severity can be assessed across treatments. This is particularly important when undertaking a complete characterization of treatments on histological, molecular, and functional outcomes, with differential effects reported across varying stroke severities [7, 8]. Additionally, it is equally important to include appropriate long-term behavioral assessments to detect neurological deficits across stroke outcomes to fully assess a potential treatment on long-term survival.

If used correctly and consistently, a model of stroke induced in the absence of anesthesia, which incorporates gradual spontaneous reperfusion, and can be used reliably in long term studies, is most desirable for the translation of new treatments to prevent the spread of injury and promote recovery in patients. Herein we describe in detail the Subheadings 2 and 3 and forms of analysis used to conduct ET-1 induced stroke in conscious rats.

2 Materials

2.1 Surgery and Stroke

This procedure uses Male Hooded Wistar rats, aged 10–12 weeks (280–360 g). Rats are housed on a 12-h day/night cycle with temperature maintained between 18 °C and 22 °C and maintained on a standard chow diet. Rats are dually housed prior to surgery, but are housed separately after surgery to ensure no disruption to the indwelling guide cannula. Be sure not to use sawdust or shredded paper in the cages as dust from these can easily settle in the ET-1 guide cannula and cause it to become blocked prior to stroke.

1. Rats.

2. Paracetamol (2 mg/kg in drinking water 24 h presurgery, postsurgery).

3. Anesthetic—Ketamine, Xylazine, Isoflurane.

4. Lignocaine (1% solution).

5. Stereotaxic frame.

6. Sterile surgical instruments including scalpel blade, curved suture needles and thread, microdissecting scissors, a fine spatula, small curved forceps.

7. 23 gauge stainless steel tubing for guide cannula (HTX-23-24).

8. 30 gauge stainless steel tubing for microinjector (HTX-30-24).

9. Small stainless steel screws (OPSM glasses frame screws).

10. Glasses frame screw driver set.

11. Hamilton glass syringe (5 μl).

12. 1 ml sterile syringe.

13. 100 μl pipette and pipette tips.

14. Clear plexiglass box.

15. Bench coat to line base of plexiglass box.

16. Premade microinjectors (*see* below).

17. Rectal thermometer.

18. Vaseline.

19. Endothelin-1 (Sigma; 40 pmol stock; dissolved in sterile saline).

20. $30^1/_2$ gauge needle.

21. Endothelin-1 infusion line: OD 0.61 × ID 0.28 mm–20 cm.

22. Cuff: OD 0.96 × ID 0.58 mm–5 mm.

23. Digital timer.

24. Fine grade metal file.

25. Heating pad.

2.2 Neurological Screening

1. Running beam (3 cm wide × 70 cm long, approx. 5 mm thick).

2. Digital Timer.

3. Small 1 cm round sticky labels in colors other than white or black.

4. Clear plexiglass box (approx. Height 30 cm, width 20 cm, length 40 cm).

5. Plexiglass cylinder (height 30 cm, diameter 20 cm, thickness 7 mm) open top.

6. Mirror (50 cm × 50 cm).

7. Video camera.

8. Rat accelerating Rota-Rod.

9. Staircase apparatus.

10. Bio-Serve Sugar pellets (45 mg each; Able Scientific, Australia).

11. Forceps for pellet positioning during training sessions.

3   Methods

3.1   Construction of Microinjector and Probe

3.1.1   Microinjector (See Fig. 1a, b)

1. To make the microinjector start with HTX-30-24 stainless steel metal tubing and cut to approximately 5 cm in length. This is best achieved by first creating a bevel along the tubing at the desired length with a small file (*see* **Note 10**) and then snapping the tubing at the bevel point to create a clean open finish. Only apply light pressure when using the file so as not to dent the lumen of the tubing. If dented ET-1 will not infuse smoothly through the injector.

2. Thread the above piece of tubing through a piece of HTX-23-24 tubing that has been cut ~2 cm in length in a similar fashion. Ensure that one end of HTX-30-24 metal tubing extends 2 cm from the end of HTX-23-24. Again use light pressure when filing as a dented tube will prevent the injector from being inserted.

3. Bend the HTX-23-24 in the middle to a 45° angle thus creating a reinforced elbow for the microinjector. Dents within the lumen of the injector can be checked by infusing distilled water smoothly through. Ensure remaining distilled water within the injector is removed before continuing.

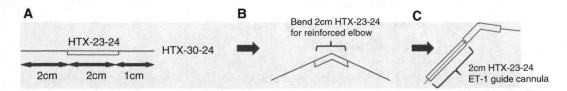

Fig. 1 Schematic diagram of the micro injector (**a**), which is threaded through a piece of 2 cm HTX-23-24 tubing to create a reinforced elbow (**b**). The final injector is later inserted into a second piece of 2 cm HTX-23-24 tubing such that it extends 2 mm from the end (**c**). This second piece of 2 cm HTX-23-24 tubing makes up the ET-1 guide cannula that is stereotaxically inserted adjacent the middle cerebral artery

3.1.2 Probe (See Fig. 1c)

1. Using a second piece of HTX-23-24, file to length so that the microinjector made above will extend 2 mm from the end of the probe once it is inserted through the probe. Hence if the microinjector measures 2 cm from the reinforced elbow then the guide probe will need to be 18 mm exactly. This 18 mm probe will be later stereotaxically positioned to sit ~3 mm dorsal to the middle cerebral artery. This distance from the artery ensures that the artery is not punctured during surgery. When the microinjector is positioned during stroke it will sit ~1 mm dorsal to the artery for endothelin-1 application. Again when filing use light pressure so as not to dent the lumen of the tubing, thus creating clean entry and exist either end of the guide cannula.

2. Place each individual microinjector into a separate clearly labeled container (glass vial) and label for each rat so that each injector made can be matched to the corresponding implanted guide probe. It is very important that each micro-injector made is an exact match to the probe implanted in the rat to ensure accurate individual localization of the middle cerebral artery during stroke induction.

3.2 Stereotaxic Surgery

The precision in placing the probe adjacent to the MCA is critical to achieving best outcomes. Attention must firstly be paid to sourcing a high quality stereotaxic frame and then ensuring no parts on the frames arm or probe holder become wobbly when set to position. Human error when reading the frame coordinates is often cause for imprecision. While some misalignment to the MCA can still result in stroke (one reason for obtaining smaller strokes), attention to detail when reading and setting coordinates is key to successful, repetitive outcomes.

1. Prior to surgery all rats should undergo physiological and neurological assessments to ensure no damage is induced as a result of stereotaxic placement of the ET-1 guide probe. This should include recording the rat's weight presurgery and assessing neurological performance according to behavioral tests described below (*see* Subheading 3.4).

2. To prevent postoperative pain, pretreat rats with Paracetamol (2 mg/kg in drinking water) 24 h prior to surgical procedure and for a further 24 h postsurgery.

3. Turn on the heating mat to support rat thermoregulation during surgery.

4. Attach the previously made guide probe (*see* above) to the probe holder on the stereotaxic frame making certain that the guide probe is positioned perfectly vertical once clamped firmly into place.

5. Anesthetize the rat by intraperitoneal injection of Ketamine (100 mg/kg)/Xylazine (10 mg/kg) solution 0.2 ml/100 g. Anesthetization is then maintained throughout surgery by inhalation Isoflurane (95% oxygen and 5% Isoflurane) via a stereotaxic nose cone attachment that features an inlet and outlet port for the flow of gaseous mixture. The oxygen gauge is set to 1.5 l and the Isoflurane gauge is set to 1 l.

6. Mount the rat into the stereotaxic frame making sure that the rat head is perfectly secure and sitting straight in the mounted position. It is important that the rat head does not wobble or move under pressure. Blunt ear probes should be positioned just above the ear canal.

7. Position the nose cone attachment on the stereotaxic frame so that it is fully covering the rat nose and apply isoflurane using an inhalation anesthetic machine as described above.

8. Once the head is secured make a mid-line incision down the center of the scalp starting just behind the eyes and moving back ~2–3 cm.

9. Place two suture threads on either side of the skin flaps and secure to each side of the frame with surgical tape, thus exposing the scalp for surgical access.

10. Gently remove the fine connective tissue layer under the scalp until the skull bone is exposed. Douse with Lignocaine to anaesthetize the surface (this will support recovery).

11. Make an incision ~ 1 cm in length along the right surface of the skull between the bone and the temporalis muscle so as to gently detach the muscle from the skull without damage to the muscle. DO NOT cut into the muscle. Use a pair of small curved forceps to pull and hold the muscle away from the skull, fixing in place with surgical tape. Clear away any residual tissue left behind on the skull bone and pat dry with a cotton bud.

12. Locate Bregma and mark with a fine tip black marker pen (*see* Fig. 2a)

13. Drill one small dent on top of the skull for placement of one small screw on the ipsilateral side close to where the ET-1 probe is destined to be positioned (*see* Fig. 2a). Gently wind in the screw such that it fixes into the skull but does not penetrate below the epidural space (about half way). This can be helped by using a set of flat forceps that hold the screw in position between the skull and screw head, thus preventing the screw from physically being pushed in all the way (*see* **Note 1**).

14. Once the screw is in place align the base of the guide probe so that it is positioned directly over and sitting on the point of bregma. Record the coordinates on the frame for Anterior/ Posterior, Medial/Lateral, and Dorsal/Ventral. Raise the

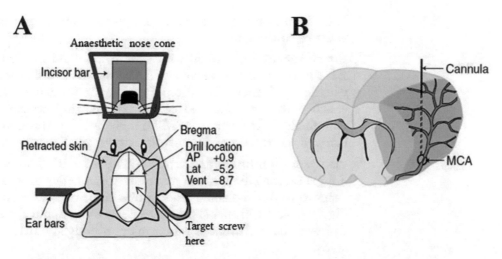

Fig. 2 Schematic diagram of stereotaxic locations: Arrows point to retraction of skin flaps, site of Bregma, site for cannula implant and positioning of anchoring screw (**a**). Diagram of the rat brain showing the target area for infusion of ET-1 adjacent to MCA (modified diagram from [9]) (**b**)

probe and adjust the frame according to the following coordinates: +0.2 AP; −5.9 ML; −5.2 DV.

15. Gently lower the guide probe until it sits on top of the point of entry into the skull. Mark this point with a black marker pen.

16. Drill a burr hole over this marked point (down the side of the skull, *see* Fig. 2a) to allow the guide probe to enter the brain and be positioned above the middle cerebral artery (*see* **Note 2**).

17. Once the probe is in position use a small amount of bone wax to seal the point of entry and secure with dental cement being sure to cover the previously attached screw to anchor the cement and probe.

18. Once the cement is dry remove the probe from the probe holder and suture wound.

19. Douse the wound with lignocaine and monitor the rat every 30 min until fully awake (approximately 2–4 h). Treat with Paracetamol in the drinking water for 24 h to prevent postoperative pain. Allow rats to recover for 4–5 days before stroke induction.

3.3 Stroke Induction

Stroke induction needs to be performed in a calm and safe environment free from disruption. The experimenter should be familiarized with each rat prior to commencing stroke. This is often best achieved while conducting prestroke physiological and behavioral assessments described below.

1. Take 60 μl of stock ET-1 (40 pmol) and add 60 μl saline to make a final 20 pmol ET-1 solution.

2. Fill a 1 ml sterile syringe with ET-1 using a 30-gauge needle attached to the syringe.

3. Attach the appropriate microinjector to one end of a 20 cm ET-1 infusion line (OD 0.61 × ID 0.28 mm) and fill the line with the endothelin-1 (20 pmol) using the 30-gauge needle attached to the other end of the line, making certain that endothelin-1 freely perfuses without air bubbles, through the line and out the tip of the microinjector as the line is filled.

4. Place a small infusion line "cuff" (OD 0.96 x ID 0.58 mm ~ 5 mm) on the elbow of the microinjector in order for it to be fastened to the implanted guide probe in the conscious rat (*see* Fig. 3a). This will enable the rat to freely move about the cage without the microinjector becoming dislodged during the study.

5. Fill a Hamilton glass syringe (5 μl) with distilled water and attach this to the infusion line, leaving a small air pocket between the endothelin-1 and the attached syringe so that ET-1 is separated from the dH$_2$O as it is pushed through. This also allows the experimenter to monitor the progression of endothelin-1 through the line by keeping track of the moving air pocket.

6. Prior to stroke use a small animal digital thermometer embalmed with Vaseline to take rectal temperature, weigh the rat, and record each for poststroke monitoring.

7. Induction of stroke itself does not cause pain. Lightly restrain the conscious rat using a soft cloth. First ensure that the external opening of the guide cannula is clear. This can be done by

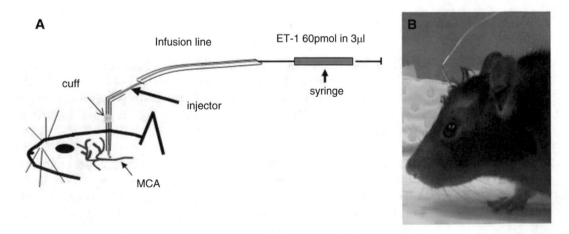

Fig. 3 Schematic diagram of the microinjector positioned into the implanted guide cannula and secured by an infusion line 'cuff' (Orange) (**a**). The external end of the microinjector is attached to the ET-1 infusion line, which is then attached to a 5 μl Hamilton syringe containing saline only. Photo of ET-1 injector correctly positioned in the conscious rat for induction of stroke (**b**)

inserting a 30 gauge needle into the tip of the cannula first (*see* **Note 12**). Then once cleared the microinjector can then be inserted into the previously implanted guide cannula, securing in place with the cuff. Be careful not to rush this process as once the microinjector is in place you do not want to bend it and compromise its final positioning within the guide probe (*see* **Note 11**).

8. Once the microinjector is positioned, place the rat in a clear plexiglass box (30 × 20 × 25 cm) for observation during stroke induction.

9. Stroke is induced by slow microinjection of ET-1 (60 pmol in 3 μl over a period of 10 min). Start timer.

10. Record all observed changes in rat behavior according to the time they occur during ET-1 infusion (*see* Table 1). It is important that during stroke rats are continually monitored and changes in normal rat behavior are observed and recorded. These behavioral changes include contralateral forepaw clenching and continuous circling (*see* Fig. 4), and can be graded and used to predict stroke outcome (*see* ref. 8). If the rats appear to be having a large stroke prior to the full 3 μl infusion, STOP INFUSION, to avoid having to euthanize the rat due to extreme stroke.

11. Monitor changes in rectal temperature every 30 min for the first 3 h after stroke. Increases in temperature (~1 °C) can occur during stroke. Temperature increases above 3 °C may result in seizure (*see* **Note 13**).

3.3.1 Monitoring Requirements

Body weight should be assessed daily throughout the duration of the study and animals continuously monitored for changes in activity. In addition to monitoring temperature and weight following stroke induction (described above) rats can also be monitored daily/weekly for neurological deficits using neurological tests specific for this species.

3.4 Assessment of Functional Outcome

When addressing the development of brain injury following stroke or traumatic brain injury it is important to measure any functional deficits that occur as a result of injury. Groups of rats can be routinely scored on a number of commonly used behavioral assessments and each rat acts as its own control preinjury. It is important to use a variety of different tests in order to pick up what are often quite subtle changes in function in rats following injury to the brain. In addition, it has been reported that rats often appear to have no deficits in one test but will still show deficits in others, depending on where the injury occurred. The behavioral tests described below will help to determine conscious limb function, fine motor control and sensorimotor function. These tests are routinely used with the ET-1 model of stroke and have been

Table 1
Changes in behavior upon ET-1 injection in the conscious rat gives an indication of stroke intensity that correlates to the histological outcomes (*see* ref. 8)

Observed behavior	Stroke rating	Stroke Infarct (unstained brain sections)
Grooming, teeth chattering Tongue poking, licking, contralateral whisker twitch, raised contralateral forepaw Raised and clenched contralateral forepaw	1	
Grooming, biting cage and bedding Spasmodic contralateral turns (not continuous) Head turned to contralateral direction Head bobbing in contralateral direction	2	
Continuous consecutive contralateral turns Chin rubbing on base of cage	3	
Contralateral forepaw clench Ipsilateral circling Forepaw shuffling/digging Tight ipsilateral circling	4	
Loss of balance on rearing/walking Loss of righting reflex Still circling after 60 min	5	

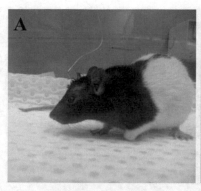
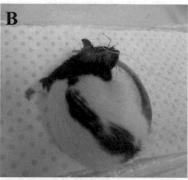

Fig. 4 Photographs of rats undergoing ET-1 induced stroke. Stroke induction is confirmed during ET-1 infusion by observed clenching of the contralateral forepaw (**a**) and in severe stroke, ipsilateral circling with clenched contralateral forepaw drag (**b**)

shown to detect immediate deficits, spontaneous recovery of deficits, and persistent long-term deficits (*see* ref. 10).

Current STAIR and RIGOUR guidelines for preclinical assessment of therapeutics to treat stroke state that all neurological assessments should be performed blinded to treatment (*see* ref. 11). Therefore, after stroke each rat should be assigned a coded number once allocated to treatment so that any further assessment is conducted blind to treatment. This generally requires two experimenters to work together, one to manage treatments and the other to perform assessments blind to treatment.

3.4.1 Neurological Deficit Score

Neurological abnormalities are evaluated with the use of a neurological deficit score based on detection of abnormal posture and hemiplegia, as described by Yamamoto and colleagues [15] (*see* Fig. 5).

1. Suspend the rat by the tail 10 cm above the bench top or home cage floor for ~5 s and observe any twisting of the thorax defined by the rat reaching up toward its tail. A nonstroke rat will extend both forepaws toward the ground as if reaching for the base of the cage (score 0) (*see* Fig. 5a). After stroke the contralateral forepaw may not reach to the ground but flex to the contralateral side. Slight flexion (score 1), 45° flexion (score 2), and pronounce 90° flexion (score 3). Often severe flexion is accompanied by obvious twisting of the thorax.

2. In addition to forelimb flexion some thorax twisting is also likely to observed in stroke affected with a wobble to contralateral side (score 1), some contralateral twisting up toward the tail (score 2), or twisting all the way up to touch the tail (score 3) (*see* Fig. 5b).

3. Set up the running beam so that it is raised approximately 20 cm above the benchtop. Limb dysfunction is detected by

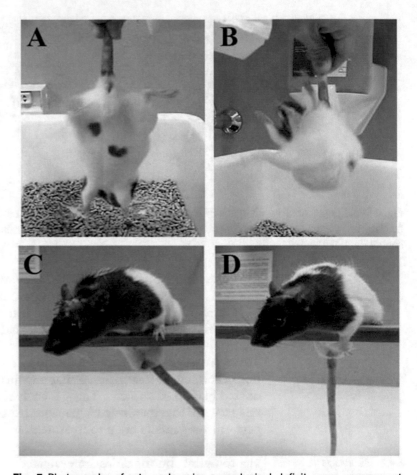

Fig. 5 Photographs of rats undergoing neurological deficit score assessment. Non-stroke affected rats reach for the cage base when suspended by the tail (**a**); Stroke affected rats contralaterally twist up toward the tail (**b**); Non-stroke affected rats grip the running beam (**c**); stroke affected rats often display limp contralateral forepaw use and hang the paw over the edge of the beam rather than grip it (**d**)

placing the rat across the narrow beam (3 cm wide × 70 cm long) and ability to grip onto the beam is observed. Forelimb and hind limb paralysis is scored by ability to grip and keep all limbs on the beam (*see* Fig. 4c). Loss of grip and occasional slipping while walking along the beam (score 1), no grip and limb resting over edge of beam (score 2) (*see* Fig. 5d), no grip and limb dangling from beam with an inability to move along the beam (score 3).

4. Add all scores from the above tests together for a total neurological deficit score with a maximum score of 12. Compare neurological deficit scores poststroke to prestroke scores such that each rat acts as its own control (*see* **Note 3**).

5. Results should be presented as box plots for nominal scores and analyzed using a Kruskal–Wallis nonparametric ANOVA followed by Dunn's posttest for multiple comparisons.

3.4.2 Sensorimotor Hemineglect (Sticky Label Test)

Sensory hemineglect is evaluated by a test developed by Schallert and Whishaw (1984) (*see* ref. 12) that measures sensitivity to simultaneous forelimb stimulation. This test is based on observations of behavior in humans with unilateral brain damage. If two stimuli are presented simultaneously, one on each side of the body, the contralateral stimulus appears to be masked ("extinguished") and either remains undetected until the ipsilateral stimulus is removed or feels subjectively weaker. In rats, the test consists of placing adhesive tapes (Avery adhesive labels, 1-cm circles) on the distal-radial region of each wrist (*see* Fig. 6).

1. Place the rat in a clear plexiglass box and allow that rat to explore the new environment for 2–3 min.

2. Gently restrain the rat and place a small (1 cm diameter) adhesive label (any color except white or black) on the inside surface of each forelimb just above the thumb, on the wrist (*see* **Note 4**). Placement of the first tape should be randomized between contralateral and ipsilateral limbs.

3. Place the rat back into the clear plexiglass box and time how long it takes for the rat to first touch each label and to remove each label. Each trial should only last a maximum of 3 min and should be conducted twice in training and then once only on subsequent days.

4. The maximum score if tape is not removed is 180 s. Compare the time to touch the tape from the contralateral forepaw to that of the ipsilateral forepaw, and plot against prestroke scores. Compare the time to remove the tape from the contralateral

Fig. 6 Photograph of a rat with adhesive labels positioned on the wrist of the forepaws for assessment of hemineglect

forepaw to that of the ipsilateral forepaw, and plot against prestroke scores.

5. Analyze using two-way repeated measures ANOVA with 2-factor repetition (side × hour after stroke) to compare latencies in the ipsilateral and contralateral forepaws over time. A one sample t-test is used to determine significance of asymmetry from chance or <0.05.

3.4.3 Rota-Rod Performance

Motor performance is commonly assessed in rodents after brain injury using an accelerating spinning wheel (Rota-Rod). While this test is a reliable measure of short term impairment, spontaneous recovery is often reported beyond 3 days and this involves learned compensatory use of the tail to maintain balance while on the wheel. As such it is not considered a reliable measure of long term motor deficits.

1. Rats are pretrained to remain on the Rota-Rod for 3 min (*see* **Note 14**). Each rat will be given two training sessions of three trials each, 1 h apart on an accelerating Rota-Rod (spinning wheel).

2. Rats are scored by timing how long they are able to remain on the Rota-Rod compared with prestroke scores.

3. When a rat falls off the Rota-Rod, it lands 20 cm below on a plastic plate which trips and stops the automatic timer. Bubble-wrap is used as a cushioning device under the Rota-Rod so that the rat does not harm itself when falling off.

4. Compare the time to fall off poststroke to that of prestroke scores such that each rat acts as its own control.

5. Analyze using a one-way ANOVA across time within groups and a two-way repeated measures ANOVA between treatment effects followed by Bonferroni post hoc test for multiple comparisons.

3.4.4 Cylinder Test

The cylinder test assesses exploratory weight bearing motor movements of rats against the wall of a cylindrical enclosure (*see* ref. 14) (*see* Fig. 7). Rats will voluntarily rear and explore the wall using their forelimbs. Prelesion rats typically distribute the weight bearing movements equally on their ipsilateral and contralateral forelimbs during vertical rearing and upon landing use both forelimbs simultaneously. After stroke, rats favor their nonimpaired limb (ipsilateral forepaw) to support them while rearing and then land on the dangling impaired limb first (*see* **Note 5**).

1. Stand cylinder on smooth clean surface.

2. Lean mirror against a wall directly behind the cylinder.

Fig. 7 Screenshots taken from video recordings of rat rearing during the cylinder test. Non-stroke affected rats will use both forepaws indiscriminately for support while rearing and exploring the cylinder (**a**); Stroke affected rats show preference toward ipsilateral forepaw use while rearing (**b**). NB: this test requires a mirror and video recording for accurate scoring of forepaw use

3. Angle the video camera in front of the cylinder so that movements by the rat on all sides of the cylinder can be seen (recommended position: 100 cm in front; 75 cm to side).

4. Start video camera before placing rat in cylinder as many movements can occur during the initial period of exploration.

5. Place the rat in the cylinder. The rat will rear and use forelimbs to explore the walls of the cylinder. Rats will typically explore walls with at least three forepaw touches before landing, therefore 10 landings should be counted in order to be confident of obtaining at least 30 vertical wall touches.

6. Repeat test 24 h after stroke, 72 h and then at 7 days with 7 day intervals thereafter (*see* **Note 6**).

7. Analyze the videotaped movements in slow motion.

8. Count only vertical wall touches where the forepaw is flat on the wall with the digits spread apart.

9. Score the first 30 vertical wall touches as either a left forelimb touch, or a right forelimb touch. Right and left forelimb movements are counted independently; if one forelimb remains stationary on the wall while the other moves, the first scores only one until it moves again.

10. Do not score ambiguous movements.

11. Compare poststroke and posttreatment scores to a prestroke baseline such that each rat acts as its own control (*see* **Note 7**).

12. Analyze using a two-way repeated-measures ANOVA followed by Bonferroni post hoc test to compare differences between treatment groups over time.

3.4.5 Staircase Test   A novel reaching test for the rat has been developed to assess the independent use of forelimbs in skilled reaching and grasping tasks that allow evaluation of sensory abilities, dexterity, and motor coordination. The apparatus is a plexiglass box with a removable baited double staircase. Food pellets are placed on the staircase and presented bilaterally at seven graded stages of reaching difficulty to provide objective measures of side bias, maximum forelimb extension and grasping skill. Animals should be well handled prior to training. The time required to train the animals for the task will take 2 weeks, with training conducted twice daily at the same time each day, taking 10–15 min each session.

The apparatus into which the animal is placed consists of a clear Perspex chamber (203 mm long × 108 mm high × 60 mm wide) with a hinged lid. A narrower compartment (165 mm long × 108 mm high × 60 mm wide) with a central raised platform running along its length, creating a 19 mm wide trough on either side, connected to the chamber. The narrowness of the side compartment prevents the animal from turning around, so that it can only use its left paw for reaching into the left trough and right paw reaching into the right trough. A removable double staircase is inserted into the end of the box, sliding into the troughs. Each of the seven steps of the staircase contains a small 3 mm deep well into which three food pellets are placed. Therefore 21 pellets are placed into the staircase. The highest step of the staircase is 13 mm below the central platform. A trained animal can collect pellets by reaching into the trough: the number of steps from which pellets have been removed provides an index of how far the rat can reach, and the number of pellets remaining at the end of the test indicates the rat's success in grasping and retrieving pellets.

Training

1. On the first day, animals are familiarized to the experimental apparatus by placing them into the test box for 15–20 min.

2. On the second day trial the rats twice. During the training period, the experimenter helps the rat into the narrower compartment. For this purpose some pellets are first distributed along the platform to attract the rats into the narrower compartment. Once in the compartment more pellets are then presented to each well with forceps in order to help the rat localize them.

3. Repeat the trials daily until the rats have learned to reach food (usually by the 4–5th training session) (*see* **Notes 8** and **9**).

Trail period: By the fifth day commence recording pellet retrieval.

1. Loading each step with three pellets for a total of 21 pellets on either side. Place rats in the test box for 15 min.

2. Count how many pellets are retrieved from each side over the 15-min trial period.

3. The final four tests will be used to determine baseline scores.

4. Only rats that can collect a minimum of 12 pellets from each side will be included in analysis of the staircase test.

5. Results are expressed as a percentage of forepaw performance compared to prestroke scores.

6. Analyze using a two-way repeated-measures ANOVA followed by Bonferroni post hoc test to compare differences between treatment groups over time

4   Notes

A number of problems can occur during surgery and stroke induction that can affect experimental outcomes. In addition to comments made above relative to each procedure, we now provide some examples with further comments regarding prevention or correction of these problems.

1. The screw is inserted first so that once the MCA coordinates are established very little pressure is then applied to the head.

2. It will be necessary to make a groove down the side of the skull to allow the probe to be inserted vertically into the brain. Test the position of drilling regularly. It is important that the guide probe is not moved out of positioned as it passes through the skull.

3. Occasionally ipsilateral deficits are also observed, particularly in the beam grip test.

4. This is different to placing the tape on the base of each forepaw. True hemineglect involves visual recognition without processing a functional response.

5. For accuracy in scoring, the test should be videotaped and analyzed in slow motion.

6. This test is not sensitive to the time period, thus exploratory movements may be encouraged by creating stimulation such as momentarily turning lights off or sliding the cylinder over a small distance. *Caution: Do not overstimulate as the rat may freeze or become agitated.*

7. Additional analysis could include horizontal landings scored as right, left, or both. The number of vertical movements per horizontal landing could then be assessed in order to determine the amount of movement the rat can accomplish per exploration.

8. Wipe the floor and walls of the starting chamber between rats, but not the staircase chamber to encourage the next rats to explore this area. Clean the test box more thoroughly at the end of each day.

9. If rats struggle to locate pellets in the wells, try raising the staircase by holding the silver handle up so that the uppermost stair is almost equal with the platform. If rats show no interest in going on to the platform try tapping the end of the staircase chamber or running a finger along the ceiling of this chamber. Poking extra pellets through the gap at the end of the staircase chamber may also encourage uncooperative rats.

10. A dull file can result in denting of the lumen when making injectors and guide cannulas, which can affect the smooth insertion of the injector into the conscious rat. This can be prevented by ensuring the file is sharp and the user has a light touch when filing.

11. When inserting the injector into the implanted guide in the conscious rat, sudden movements while pushing it through can result in bending of the injector. This renders the injector unusable. We therefore recommend that each injector and guide combination is documented for exact length such that a new injector could be made to fit the implanted guide.

12. Occasionally an implanted guide cannula can become blocked (often due to dust or small blot clots as a result of surgery). We recommend using a 30 gauge needle to clear the external end of the guide cannula prior to inserting the injector. If the block appears to be occurring closer to the internal end of the cannula, try infusing a small amount of saline first in an attempt to dissolve the block. If this does not work this rat could be used in a sham surgical group.

13. There are a number of expected risks associated with stroke induction. Temperature increases above 3 °C during stroke may result in seizure and rats displaying these signs should be humanely euthanized immediately (Lethobarbitone 1:2 dilution, ~160 mg/kg i.p.). Any rat showing signs of prolonged loss of righting reflex or seizure activity during or after stroke should also be humanely euthanized (Pentobarbitone 1:2 dilution, ~160 mg/kg i.p.). Body weight should be assessed daily throughout the duration of the study: loss of weight greater than 20% of preweight stroke should be keenly assessed for euthanasia. Rats experiencing a severe stroke may lose some forelimb/hind limb dexterity but stroke does not usually result in complete loss of function.

14. During Rota-Rod training many rats learn to turn around while on the wheel, enabling them to drop off quickly. This

can be avoided during training by taping the nose of the rat with a pen tip each time the rat's head tries to turn around.

References

1. Yanagisawa M et al (1988) A novel potent vasoconstrictor peptide produced by vascular endothelial cells. Nature 332:411–415

2. Haynes WG, Webb DJ (1998) Endothelin as a regulator of cardiovascular function in health and disease. J Hypertens 16:1081–1098

3. Sharkey J et al (1993) Perivascular microapplication of endothelin-1: a new model of focal cerebral ischaemia in the rat. J Cereb Blood Flow Metab 13:865–871

4. Virley D et al (2004) A new primate model of focal stroke: endothelin-1-induced middle cerebral artery occlusion and reperfusion in the common marmoset. J Cereb Blood Flow Metab 24:24–41

5. Biernaskie J et al (2001) A serial MR study of cerebral blood flow changes and lesion development following endothelin-1-induced ischemia in rats. Magn Reson Med 46:827–830

6. Counsell C et al (2002) Predicting outcome after acute and subacute stroke: development and validation of new prognostic models. Stroke 33:1041–1047

7. Roulston C et al (2012) Animal models of stroke for preclinical drug development: a comparative study of flavonols for cytoprotection. Translational stroke research: from target selection to clinical trials. In: Lapchak P, Zhang J (eds) Translational stroke research. Springer Publishing, New York. Chapter 25, pp 493–524

8. Roulston CL et al (2008) Using behaviour to predict stroke severity in conscious rats: post-stroke treatment with 3′, 4′-dihydroxyflavonol improves recovery. Eur J Pharmacol 584:100–110

9. O'Neil MJ, Clemens JA (2001) Rodent models of focal cerebral ischemia. Curr Protoc Neurosci. Chapter 9: Unit 9.6

10. Schaar KL et al (2010) Functional assessments in the rodent stroke model. Exp Transl Stroke Med 2:13

11. Lapchak PA et al (2013) RIGOR guidelines: escalating STAIR and STEPS for effective translational research. Transl Stroke Res 4:279–285

12. Schallert T, Whishaw IQ (1984) Bilateral cutaneous stimulation of the somatosensory system in hemidecorticate rats. Behav Neurosci 98:518–540

13. Fuxe K et al (1992) Involvement of local ischemia in endothelin-1 induced lesions of the neostriatum of the anaesthetized rat. Exp Brain Res 88:131–139

14. Soleman S et al (2010) Sustained sensorimotor impairments after endothelin-1 induced focal cerebral ischemia (stroke) in aged rats. Exp Neurol 222:13–24

15. Yamamoto M et al (1988) Behavioral changes after focal cerebral ischemia by left middle cerebral artery occlusion in rats. Brain Res 452:323–328

Chapter 11

A Murine Model of Hind Limb Ischemia to Study Angiogenesis and Arteriogenesis

Jun Yu and Alan Dardik

Abstract

Therapeutic angiogenesis offers promise as a novel treatment that is complementary to surgical or endovascular procedures for peripheral arterial diseases (PAD). Appropriate development and use of hind limb ischemia models is necessary for successful studies of therapeutic angiogenesis and/or arteriogenesis. In this chapter, we describe two commonly used murine unilateral hind limb ischemia models, the femoral artery transection model and the femoral/saphenous artery excision model.

Key words Hind limb ischemia, Animal model, Angiogenesis, Arteriogenesis, Ischemia, Peripheral arterial disease

1 Introduction

Peripheral arterial disease (PAD) affects approximately eight million people in the USA, although it has poor awareness and recognition in the general population, and thus the prevalence of PAD is likely to be underestimated [1]. While surgical revascularization remains the most effective treatment for limb ischemia, many patients with advanced disease are not suitable for surgical or endovascular management. Therapeutic angiogenesis and stem cell therapies are newer treatments that may be appropriate for these difficult patients [2, 3]. However, defining optimal parameters for these gene and cell therapies is expensive to perform as clinical trials in human patients, and therefore may be more suitable in animal models [4, 5]. The murine hind limb ischemia model is a very useful model to use for testing some of these parameters and therapies [5, 6].

In this chapter, we describe the methodology for the two most commonly used murine models of unilateral hind limb ischemia, e.g., the femoral artery transection model and the femoral/saphenous artery excision model. The femoral artery transection model is a model that produces only a mild-to-moderate amount of ischemia

Binu Tharakan (ed.), *Traumatic and Ischemic Injury: Methods and Protocols*, Methods in Molecular Biology, vol. 1717,
https://doi.org/10.1007/978-1-4939-7526-6_11, © Springer Science+Business Media, LLC 2018

that induces mainly arteriogenesis in the thigh, with minimal calf angiogenesis. The femoral/saphenous artery excision model produces more severe ischemia that induces both thigh arteriogenesis as well as calf angiogenesis. The materials and procedures for establishing and evaluating these models are described, including several notes that come out of our own experience with these models.

2   Materials

2.1   Surgical Tools (See Note 1)

1. Dissecting microscope.
2. Pointed forceps.
3. Surgical scissors.
4. Retractor (*see* **Note 2**).
5. Fine pointed forceps.
6. Spring scissors.
7. Needle holder.
8. Cautery tool.
9. Sterile cotton swabs.
10. 7-0 and 6-0 nonabsorbable sutures.
11. Heating pad.

2.2   Blood Flow Measurement Equipment

1. PeriFlux Laser Doppler Perfusion Measurement Unit with deep penetration probe.
2. Laser speckle imaging system (moorFLPI-2 or higher resolution moorLDI2-HIR).
3. Power lab.

3   Methods

3.1   Femoral Artery Transection Model (See Note 3)

1. Anesthetize the mouse by placing it into an anesthesia induction chamber containing 2% isoflurane at a flow rate of 2 L/min.
2. Remove the mouse from the induction chamber when it is unresponsive to external stimuli. Confirm proper anesthetization by pinching its toe. Apply some artificial tears ointment on the eyes to prevent dryness (*see* **Note 4**).
3. Place the mouse in the preoperating area and connect it to a continuous flow of isoflurane. Remove the hair from the lower abdomen to the foot using an electric shaver. Then apply hair removal cream to thoroughly remove the fur.

4. Place the mouse in a supine position over a heating pad with the left hind limb slightly abducted and the knee joint slightly flexed. Extend and secure the limbs with a piece of tape with hind feet facing up. Connect the mouse to a continuous flow of isoflurane.

5. Once the mouse is secure, prepare the skin around surgical area with three alternating betadine and alcohol scrubs.

6. Make a ~1 cm midline incision of the skin from ~7 mm below the inguinal region and ~3 mm above it using forceps and surgical scissors. The remainder of the surgical procedure should be performed under a dissecting microscope to gain a magnified view of the surgical region.

7. Gently push away subcutaneous fat tissue superficial to the neurovascular bundle in the thigh region by using phosphate buffered saline moistened cotton swabs and forceps.

8. Use a retractor to open the wound for easy access to the neurovascular bundle.

9. Gently pierce through the membrane of femoral sheath by blunt dissection using fine forceps to expose the neurovascular bundle. The anatomy of the surgical region should be now clearly exposed as shown in (Fig. 1). Asterisks indicate the locations of ligation for the induction of hind limb ischemia.

10. Push away the nerve from the femoral artery and vein by using a wet cotton swab.

11. Carefully separate the femoral artery from the femoral vein at the ligation sites between the proximal caudal femoral artery and the bifurcation of the deep femoral artery and saphenous artery by blunt dissection using surgical fine forceps. Use caution not to tear the femoral vein (*see* **Note 5**).

12. Then, pass two 7-0 silk sutures underneath the separated femoral artery segment. Tie off the artery by double knots as shown in Fig. 2a.

13. Transect the femoral artery between the two ties.

14. If a Laser Doppler Perfusion system is used, the blood flow should be assessed at this point, as described below (*see* Subheading 3.3).

15. Remove the retractor and close the incision using 6-0 Nylon sutures.

16. If Laser Doppler imaging system is used, the blood flow should be assessed at this point, as described below (*see* Subheading 3.4).

17. After blood flow measurement, place the mouse in a clean cage with a heating pad and monitor the breath and heart beat continuously until the animal is fully recovered.

18. Return the mouse to animal facility (*see* **Note 6**).

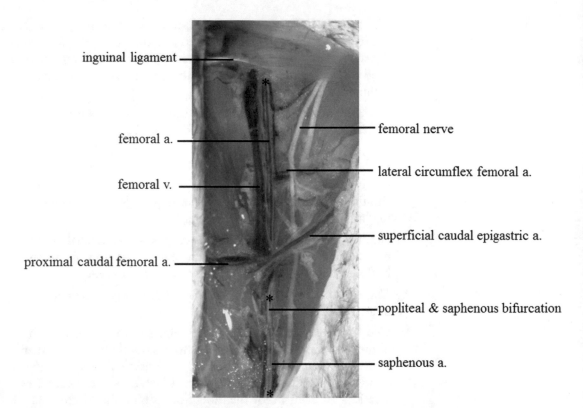

inguinal ligament

*

femoral a.

femoral v.

proximal caudal femoral a.

*

femoral nerve

lateral circumflex femoral a.

superficial caudal epigastric a.

popliteal & saphenous bifurcation

saphenous a.

*

Fig. 1 Anatomy of the mouse hind limb (left). Asterisks indicate the locations of ligation in the models of hind limb ischemia. A indicate artery; v indicate vein

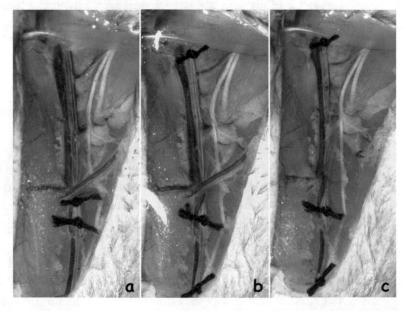

Fig. 2 (**a**) The femoral artery transection model. Two ligatures around the femoral artery (not the vein), before transection of the intervening segment. (**b**) The femoral/saphenous artery excision model. Upper ligature around the femoral artery, middle ligature around the popliteal/saphenous artery bifurcation, and the lower ligature around the saphenous artery. (**c**) The femoral/saphenous artery excision model, after excision of the femoral and saphenous arteries

3.2 Femoral Artery and Saphenous Artery Excision Model (See Note 7)

1. Anesthetize the mouse by placing it into an anesthesia induction chamber containing 2% isoflurane at a flow rate of 2 L/min.

2. Remove the mouse from the induction chamber when it is unresponsive to external stimuli. Confirm proper anesthetization by pinching its toe. Apply some artificial tears ointment on the eyes to prevent dryness (*see* **Note 4**).

3. Place the mouse in the preoperating area and connect it to a continuous flow of isoflurane. Remove the hair from the lower abdomen to the foot using an electric shaver. Then apply hair removal cream to thoroughly remove the fur.

4. Place the mouse in a supine position over a heating pad with the left hind limb slightly abducted and the knee joint slightly flexed. Extend and secure the limbs with a piece of tape with hind feet facing up. Connect the mouse to a continuous flow of isoflurane.

5. Once the mouse is secure, prepare the skin around surgical area with three alternating betadine and alcohol scrubs.

6. Make a ~1 cm midline incision of the skin from ~7 mm below the inguinal region and ~3 mm above it using forceps and surgical scissors. The remainder of the surgical procedure should be performed under a dissecting microscope to gain a magnified view of the surgical region.

7. Gently push away subcutaneous fat tissue superficial to the neurovascular bundle in the thigh region by using phosphate buffered saline moistened cotton swabs and forceps.

8. Use a retractor to open the wound for easy access to the neurovascular bundle.

9. Gently pierce through the membrane of femoral sheath by blunt dissection using fine forceps to expose the neurovascular bundle. The anatomy of the surgical region should be now clearly exposed as shown in (Fig. 1). Asterisks indicate the locations of ligation for the induction of hind limb ischemia.

10. Push away the nerve from the femoral artery and vein by using wet cotton swab.

11. Carefully separate the femoral artery from the femoral vein at the proximal ligation site just below the inguinal ligament. Use caution not to tear the femoral vein (*see* **Note 5**).

12. Then, pass a 7-0 silk suture underneath the proximal end of the femoral artery just below the inguinal ligament and tie off the artery by double knots as shown in (Fig. 2a).

13. Carefully separate the femoral artery from the femoral vein all the way down to the site of popliteal artery and saphenous

artery bifurcation by blunt dissection using surgical fine forceps. Use caution not to tear the femoral vein.

14. Then, pass a 7-0 suture underneath the femoral artery just proximal to the popliteal artery and saphenous artery bifurcation. Tie off the artery by double knots as shown in (Fig. 2b).

15. Replace the retractor if necessary so the distal region of femoral artery and saphenous artery can be exposed clearly (*see* **Note 8**).

16. Carefully separate the saphenous artery from the vein for about half way down by blunt dissection using surgical fine forceps. Use caution not to tear the saphenous vein.

17. Then, pass a 7-0 suture underneath the distal site of saphenous and tie off the artery by double knots as shown in (Fig. 2b).

18. Using a cautery tool to cauterize the lateral circumflex femoral artery and proximal caudal femoral artery. Apply the cautery transversely to incise the superficial caudal epigastric artery.

19. Transect and excise the segment of artery between the most distal and proximal knots by using a pair of spring scissors. Use caution not to tear the surrounding vein (Fig. 2c).

20. If a Laser Doppler Perfusion system is used, the blood flow should be assessed at this point, as described below (*see* Subheading 3.3).

21. Remove the retractor and close the incision using 6-0 Nylon sutures.

22. If Laser Doppler imaging system is used, the blood flow should be assessed at this point, as described below (*see* Subheading 3.4).

23. After blood flow measurement, place the mouse in a clean cage with a heating pad and monitor the breath and heart beat continuously until the animal is fully recovered.

24. Return the mouse to animal facility (*see* **Note 6**).

3.3 Laser Doppler Perfusion Measurement

1. Place the mouse into the anesthesia induction chamber containing 2% isoflurane at a flow rate of 2 L/min.

2. After the mouse is unresponsive to external stimuli, remove it from the induction chamber. Confirm proper anesthetization by pinching its toe. Apply some artificial tears ointment on the eyes to prevent dryness.

3. Place the mouse on preoperating region connected to isoflurane.

4. Apply hair removal cream to thoroughly remove the fur below knee.

5. Place the animal on a 37 °C heated surface for 5 min. Monitor the body temperature to ensure stable.

6. Place the mouse in a supine position over a heating pad with the both hind limbs slightly abducted. Extend and secure the limbs with a piece of tape with paws facing up.

7. Once the mouse is secure, prepare the skin below knee area with three alternating betadine and alcohol scrubs.

8. Make a ~7 mm midline incision of the skin below knee on both limbs.

9. Place the deep penetration probe onto the surface of gastrocnemius muscle group.

10. Record the blood flow trace for 3 min for each limb (*see* **Note 9**).

11. Close the incision using 6-0 Nylon sutures.

12. Return the mouse to the recovery cage and monitor the animal continuously until fully recovered.

3.4 Laser Doppler Imaging

1. Place the mouse into the anesthesia induction chamber containing 2% isoflurane at a flow rate of 2 L/min.

2. After the mouse is unresponsive to external stimuli, remove it from the induction chamber. Confirm proper anesthetization by pinching its toe. Apply some artificial tears ointment on the eyes to prevent dryness.

3. Place the mouse in the preoperating area and connect it to a continuous flow of isoflurane.

4. Apply hair removal cream to thoroughly remove the fur below knee.

5. Place the animal on a 37 °C heated surface for 5 min. Monitor the body temperature to ensure stable.

6. Place the animal in the supine position on a nonreflective light-absorbing surface, connected to a continuous flow of isoflurane. Extend the hind limbs.

7. Specify the size of the field of view and scanning resolution (*see* **Note 10**). Using the Laser Doppler imager and the acquisition module to acquire image per the manufacturer's instructions (*see* **Note 9**).

8. Return the mouse to the recovery cage and monitor the animal continuously until fully recovered.

Both Laser Doppler Perfusion Measurement and Laser Doppler imaging procedures can be repeated to follow the changes in blood flow recovery over time.

4   Notes

1. Surgical tools need to be sterilized prior to surgery with a hot-bead sterilizer. It is required that the surgical tools be resterilized between each operation when multiple mice are operated on the same day.

2. We use our own retractors since most commercial available retractors are too big for surgery on a mouse limb.

3. This is a mild hind limb ischemia model that mainly induces arteriogenesis in the thigh without calf angiogenesis.

4. Flush the anesthetic from the induction chamber prior to opening the lid to decrease the operator's exposure to isoflurane.

5. In the case of accidental disruption of the femoral vein or other vessels occur, a dry sterile cotton tipped applicator should be applied to the site of hemorrhage with moderate pressure until the bleeding stops.

6. Consult the Institutional Animal Care and Use Committee (IACUC) at your institution for specific regulations regarding returning post-surgical animals to the animal care facility, as the mouse that undergoes hind limb ischemia surgery may be considered to have a pain level of either D or E, depending on the exact protocol and medications used. Appropriate pre-operation and postoperation analgesic agents should be given per local IACUC regulations.

7. This is a more severe hind limb ischemia model that induces arteriogenesis in the thigh and profound hypoxia induced angiogenesis in the calf muscle groups.

8. The mouse skin on the lower limb is very flexible. It is possible to retract the wound up and down to expose the proximal and distal surgical regions by using a smaller incision. In our experience, this procedure can accelerate the wound closure time, decrease the inflammatory response cause by surgery and diminish the impact of surgical wound on Laser Doppler imaging. A larger incision is shown in the Figures only for illustrative purposes.

9. The body temperature and the light in the measurement area should be kept very stable. Any vibration should be avoided. All these factors may influence the recording.

10. The field of view and scanning resolution should be consistent between animals and all the measurements at different time points.

Acknowledgment

This work was supported by NIH grants R01HL126933 and R56HL117064 and a grant from American Heart Association 14GRNT20450093 to JY; and NIH grants R01-HL095498 and R56-HL095498 as well as the United States Department of Veterans Affairs Biomedical Laboratory Research and Development Program Merit Review Award I01-BX002336 to AD.

References

1. Roger VL, Go AS, Lloyd-Jones DM et al (2011) Heart disease and stroke statistics update: a report from the American Heart Association. Circulation 123:e18–e209

2. Bautch VL (2011) Stem cells and the vasculature. Nat Med 17:1437–1443

3. Powell RJ, Marston WA, Berceli SA, Guzman R, Henry TD, Longcore AT, Stern TP, Watling S, Bartel RL (2012) Cellular therapy with Ixmyelocel-T to treat critical limb ischemia: the randomized, double-blind, placebo-controlled RESTORE-CLI trial. Mol Ther 20:1280–1286

4. Brenes RA, Bear M, Jadlowiec C, Goodwin M, Hashim P, Protack CD, Ziegler KR, Li X, Model LS, Lv W, Collins MJ, Dardik A (2012) Cell-based interventions for therapeutic angiogenesis: review of potential cell sources. Vascular 20:360–368

5. Brenes RA, Jadlowiec CC, Bear M, Hashim P, Protack CD, Li X, Lv W, Collins MJ, Dardik A (2012) Toward a mouse model of hind limb ischemia to test therapeutic angiogenesis. J Vasc Surg 56:1669–1679

6. Limbourg A, Korff T, Napp LC, Schaper W, Drexler H, Limbourg FP (2009) Evaluation of postnatal arteriogenesis and angiogenesis in a mouse model of hind-limb ischemia. Nat Protoc 4:1737–1746

Chapter 12

A Murine Model of Myocardial Ischemia–Reperfusion Injury

Zhaobin Xu, Kevin E. McElhanon, Eric X. Beck, and Noah Weisleder

Abstract

Ligation of the left anterior descending (LAD) coronary artery in the mouse heart is a widely used model to simulate myocardial infarction and ischemia–reperfusion injury. Here we describe a ligation technique routinely performed in our laboratory to induce myocardial infarction that may be used to study ischemia–reperfusion injury in the myocardium. The methods described enhance location of the LAD coronary artery to allow for accurate ligation, thus increasing reproducibility of infarct size and location.

Key words Myocardial ischemia–reperfusion, Left anterior descending coronary artery, Myocardial infarction, LAD ligation, Permanent ligation

1 Introduction

Coronary heart disease, despite a steady decline in the rate of deaths, is responsible for 1 of every 6 deaths in the USA [1]. This fact illustrates the importance of animal models of myocardial infarction (MI) and ischemia/reperfusion (I/R) injury to facilitate our understanding of the pathophysiology underlying myocardial damage resulting from coronary heart disease [2]. The ligation of coronary arteries has long been utilized to understand changes in the electrophysiology and cardiac function in large animal models [3–6], however, advances in the field of transgenic mouse models offer the opportunity for deeper exploration of cardiac pathophysiology at the molecular level. Ligation of the left anterior descending (LAD) coronary artery in a murine model to simulate the ischemic events associated with MI was established in 1995 by Michael et al. [7]. Early surgical techniques of LAD ligation in mice necessitated major invasive dissection of the thoracic cavity resulting in complications during surgery [8] and difficulty in accurate placement of the LAD ligature compromising reproducibility [9, 10].

Zhaobin Xu and Kevin E. McElhanon contributed equally to this work.

Binu Tharakan (ed.), *Traumatic and Ischemic Injury: Methods and Protocols*, Methods in Molecular Biology, vol. 1717, https://doi.org/10.1007/978-1-4939-7526-6_12, © Springer Science+Business Media, LLC 2018

The methods described in this chapter are adapted from multiple previously described LAD ligation techniques [11–14]. In this procedure, the mouse is anesthetized under controlled conditions and placed on a ventilator, remaining so for the duration of the ischemic period. The thoracic cavity is opened with a 6–8 mm incision at the level of the third intercostal space and ligation of the left anterior descending coronary artery is performed. The chest is temporarily closed for the desired period of ischemia to prevent dehydration and to maintain a clean surgical site. After the ischemic period, the thoracic cavity is again opened and the ligation removed returning normal blood flow to the ischemic myocardium of the left ventricle. Alternatively, the ligature can remain in place to create a permanent occlusion of blood flow. The chest cavity is then sutured and excess air removed from the thoracic cavity by applying gentle pressure. Once the surgical site is closed, an analgesic is administered and anesthesia is ceased, however the mouse remains on the ventilator. Once the mouse begins to exhibit righting reflex the ventilator may be turned off and the trachea tube removed.

Utilizing our described method, the flow of blood supplying the myocardium of the left ventricle can be permanently blocked or the ligature removed to allow blood to reperfuse the previously ischemic myocardium to model coronary blockage that is later cleared spontaneously or through medical intervention. Because this method effectively and reproducibly mimics myocardial infarction and ischemia-reperfusion injury, it may be utilized in studies of prophylactic or therapeutic drugs, histological studies of inflammatory cell infiltration and postinfarction tissue remodeling, as well as molecular responses and protein regulation in the injured heart. While the described methods are in wide use in the field of cardiovascular research, new methodologies with distinct advantages are gaining acceptance. For example, the technique described by Gao et al. [15] eliminates the need for intubation and reduces tissue damage and associated inflammation, leading to increasing use of this approach. Regardless of the specific surgical approach, the increasing number of readily available genetically modified mice means that the murine model of LAD coronary artery ligation will remain an invaluable surgery for cardiovascular research [16, 17].

2   Materials

Throughout the surgical procedure it is imperative to maintain a clean operating field. Standard procedures for animal survival surgery should be observed, including autoclaving all instruments and surgical supplies before use. Sterile, single-use surgical gloves and proper gowning of the surgeon should be used during the procedure.

2.1 Surgical Equipment	1. PhysioSuite with RightTemp Homeothermic Warming System (Kent Scientific).
	2. Gooseneck light source (Zeiss).
	3. Omano trinocular stereoscope (Microscope.com).
	4. SB2 boom stand with universal arm (Micrcoscope.com).
	5. Isoflurane anesthesia systems for rodents and small animals (VetEquip).
	6. Medical grade oxygen (Cylinder Gas).
	7. MiniVent type 845 respirator (Harvard Apparatus).

2.1 Surgical Equipment

1. PhysioSuite with RightTemp Homeothermic Warming System (Kent Scientific).
2. Gooseneck light source (Zeiss).
3. Omano trinocular stereoscope (Microscope.com).
4. SB2 boom stand with universal arm (Micrcoscope.com).
5. Isoflurane anesthesia systems for rodents and small animals (VetEquip).
6. Medical grade oxygen (Cylinder Gas).
7. MiniVent type 845 respirator (Harvard Apparatus).

2.2 Anesthesia and Analgesia

1. Isoflurane.
2. Buprenorphine, stock 0.3 mg/mL, dilute with 0.9% sterile saline to 0.1 mg/mL, administer 100 μL per mouse, subcutaneously.
3. 0.9% sterile saline.
4. Bupivacaine, local anesthetic for surgical site.
5. Ibuprofen (Motrin).

2.3 Surgical Tools

1. Fine scissors (Fine Science Tools, FST).
2. #5 Dumont forceps (FST).
3. #3 Dumont forceps (FST).
4. Castroviejo microneedle holders (FST).
5. Slim elongated needle holder (FST).
6. Round handled needle holder (FST).
7. Micro locking forceps (FST).
8. Chest retractor (FST).
9. Bulldog clamp (FST).
10. #11 scalpel (B-D Bard Parker).
11. 4-0 silk suture, taper needle (Sharpoint Products).
12. 6-0 silk suture, taper needle (Sharpoint Products).
13. Y connector to tracheal tube, 0.5 mm, 1/16 in., (Kent Scientific).

2.4 Miscellaneous Sterile Supplies

1. Betadine Surgical Scrub (Purdue Products).
2. 70% ethanol (Fisher Scientific).
3. Laboratory or surgical tape (Fisher Scientific).
4. Sterile gauze pads (Fisher Scientific).
5. Sterile surgical drape (Fisher Scientific).
6. Single use sterile surgical gloves.
7. Angiocatheter, 20 gauge, used as an intubation tube (Henry Schein).

3   Methods

3.1   Preparation for Surgery

1. At least 12 h prior to the initiation of surgical procedures the animal's drinking water should be supplemented with 0.2 mg/mL Ibuprofen (resulting in a dose of 30 mg/kg) to allow for prophylactic prevention of pain and distress in animals.

2. Prepare the surgical surface by covering with sterile drape (*see* **Note 1**). Use tape to secure a loop of 4-0 silk suture to the surgical surface near where the head of the mouse will be placed so that when the mouse is positioned for intubation the loop of 4-0 suture can be placed around the upper incisors of the mouse.

3.2   Shaving and Anesthesia (This Procedure Assumes Isoflurane as an Anesthetic. A Ventilator Is Still Required If Using Ketamine/Xylazine or Other Anesthesia.)

1. The incision sites on the mouse's chest and neck should be shaved using an electric hair shaver at a site away from the surgical surface to maintain a sterile field. Remove any remaining loose hair by wiping or briefly vacuuming.

2. Initial anesthesia is induced by placing the mouse into an induction chamber filled with 5% isoflurane and 0.4 L/min oxygen. The mouse remains in the chamber until there is a clear loss of its righting reflex. The tail will also lose tone.

3. Transfer the mouse to the surgery surface so that the mouse is on the homeothermic heating pad in a supine position with its head facing the surgeon.

4. Apply a nose cone that supplies 2% isoflurane, 0.4 L/min oxygen to maintain anesthesia. Deep anesthesia must be accomplished before continuing with the surgery, which can be tested by checking for a response to a toe pinch. In order to produce deep anesthesia the levels of isoflurane provided through the nose cone may be adjusted as needed (*see* **Note 2**).

3.3   Positioning of Mouse for Surgery

1. Secure the mouse incisors with the suture loop from above (Subheading 3.1, **step 2**) so that the mouth is pulled slightly open to allow for subsequent intubation.

2. Pull the tail of the mouse until the body is straight and then secure the tail to the surgical surface with tape. The mouse body should be taut, however, avoid stretching.

3. The forelimbs can be spread out and attached to the surgical surface with tape. Again, avoid overstretching the limbs as pulling on the thorax can have effects on respiration during the remainder of the surgery.

4. Turn on the homeothermic heating pad. This will help to maintain the body temperature of the mouse throughout the surgery.

3.4 Endotracheal Intubation

1. Prepare a 20-gauge angiocatheter for use as an intubation tube by removing the inner needle.

2. Prepare the shaved neck for a surgical incision by three scrub cycles. Each cycle includes first swabbing with betadine and then cleaning the site with 70% ethanol.

3. Inject bupivacaine (50 μL of 0.1% stock) intradermally at the incision site to provide local analgesic.

4. Expose the trachea and thyroid gland by making a 0.5 cm long incision through the skin layer.

5. Visualize the trachea under the sternohyoideus muscle by carefully separating the lobes of the thyroid gland at the isthmus.

6. Use blunt forceps to move the tongue up toward the lower jaw and to one side with one hand and then insert the 20-gauge angiocatheter into the mouth using the other hand.

7. Gently insert the 20-gauge angiocatheter into the trachea while observing it through the incision. If the angiocatheter cannot be observed entering the trachea it has been inserted into the esophagus. In this case the angiocatheter should be removed and the procedure repeated (*see* **Notes 3** and **4**).

3.5 Maintenance of Ventilation

1. Ventilation of the mouse is maintained by the MiniVent TYPE 845 (Harvard Apparatus) using room air supplemented by isoflurane.

2. For a 25 g mouse a tidal volume of approximately 260 μL/stroke can be used. The tidal volume should be adjusted according to body weight, with higher body weights requiring an increased tidal volume. Generally, a ventilation rate of 110 strokes/min is sufficient.

3. To confirm that the angiocatheter is properly inserted into the trachea and proper ventilation is achieved, it is necessary to observe symmetrical expansion of the thorax. Uniform bilateral chest expansion indicates that the angiocatheter is properly inserted.

3.6 Thoracotomy

1. Reposition the mouse for the thoracic incision by removing the tape on the tail and left forelimb and then turning the mouse into a right lateral decubitus position. Once the mouse is turned the limbs and tail can be taped to the surgical surface once again, taking care not to overstretch the limbs and pull excessively on the thorax.

2. Prepare the surgical incision site by three scrub cycles. Each cycle includes first swabbing with betadine and then cleaning the site with 70% ethanol.

3. Inject bupivacaine (50 μL of 0.1% stock) intradermally at the incision site.

4. Make an oblique 1 cm incision approximately 2 mm away from the left sternal border and approximately 1–2 mm toward the junction of the left front leg with the body (*see* **Note 5**).

5. Establish the position of the superficial thoracic vein at the lateral corner of the incision.

6. Avoid the superficial thoracic vein when making the incision through the layers of muscle on the thorax (*see* **Note 6**). This incision exposes the ribs.

7. Use scissors to make a small (6–8 mm) incision into the thorax cavity at the third intercostal space approximately 2–3 mm from the left sternal border (*see* **Note 7**), being careful to avoid any contact with the lungs.

8. Spread the incision by inserting the chest retractors and then gently apply pressure until the incision opens to approximately 8–10 mm in width, which should expose the heart and lungs.

9. Use a pair of curved forceps to gently lift and then separate the pericardium. The separated edges of the pericardium can be gently inserted into the retractor with the curved and straight forceps by placing the edges into the retractor to remove them from the surgical field.

10. Identify the LAD coronary artery on the surface of the heart. It appears as a thin red line within the myocardial wall originating from the left main coronary artery running perpendicular from the left atrium and longitudinally toward the apex of the heart. A direct, focused light on the area may assist in identification.

3.7 LAD Ligation

1. Locate the position on the LAD to be ligated. The usual ligation site is 1–2 mm below the left auricle, a position where the ligature will produce ischemia in roughly 40–50% of the left ventricle. A more distal ligation site will produce a smaller infarct zone.

2. Once the ligation site has been determined, gently press the artery with the curved forceps slightly distal to the subsequent ligation point to trap blood and temporarily enlarge the LAD. This maneuver can be repeated to assist in identification of the LAD.

3. Under the dissecting microscope, use a tapered needle to pass a 6-0 silk ligature through the myocardial wall muscle underneath the LAD (*see* **Note 8**).

4. Tie the 6-0 silk ligature with a loose double knot that has a 2–3 mm diameter loop.

5. Insert a 2–3 mm long piece of PE-10 tubing into the loop parallel to the artery. Proper positioning of the tubing can require several attempts.

6. Gently tighten the loop of the 6-0 silk ligature until it is tight around the artery and tubing then secure the ligature in place with a slipknot. Excessive pressure when tightening the ligature can damage the myocardial wall.

7. Confirm the cessation of blood flow through the LAD by observing a color change in the anterior wall of the left ventricle. Occlusion of the LAD causes it to become a paler color because of the reduced blood perfusion. If permanent ligation of the LAD is required, remove the PE-10 tubing and the remainder of the procedure can be resumed at Subheading 3.9.

8. Remove the retractors from the incision and then inflate the lungs by blocking the ventilator outflow for 1–2 s. The mouse remains on the ventilator for the duration of the LAD occlusion.

9. Close the incision temporarily with a bulldog clamp during the ischemic period. The length of time for this ischemic period depends on the particular experimental design, but is usually between 20 and 60 min.

3.8 Reperfusion

1. Remove the bulldog clamp once the ischemic period ends and insert the retractors to open the incision and expose the thoracic cavity.

2. Untie the slipknot in the 6-0 silk ligature and then remove the PE-10 tubing. This should allow reperfusion of the ischemic tissue, which can be confirmed by a color change back to the normal red color of the left ventricle within 20 s.

If analysis of the infarct zone by triphenyl tetrazolium chloride (TTC) and/or phthalo blue staining (to identify the at risk zone) will be performed after the reperfusion period the 6-0 silk ligature should be loosened but left in place. Otherwise, the 6-0 ligature can now be removed.

3.9 Closure of Incision

1. Remove the retractors and close the thorax incision by suturing together the third and fourth ribs with a single 4-0 silk suture using care to avoid touching the lungs (*see* **Note 9**).

2. Use 4-0 silk as a continuous suture to separately close the muscle and skin layers.

3. Cease the flow of isoflurane while maintaining the flow of room air at 0.4 L/min.

3.10 Postoperative Care

1. Monitor the mouse for signs of recovery from anesthesia, including movement of the tail or whiskers (*see* **Note 2**) which will be followed by attempts at spontaneous respiration.

2. Once the mouse resumes a normal breathing pattern, extubate the mouse by removing the angiocatheter slowly to avoid aspiration into the lungs.

3. Allow the mouse to recover for an additional 3–5 min to confirm that the mouse is not experiencing respiratory distress.

4. Once the mouse begins to breathe on its own, inject 100 µL of 0.1 mg/mL buprenorphine subcutaneously. The mouse should be monitored post-op and an injection of buprenorphine should be provided every 6–12 h over the next 72 h.

5. Ibuprofen is also provided in the drinking water as a 0.2 mg/mL solution for additional pain relief for up to a maximum of 7 days after surgery.

6. Consult with local veterinary staff to minimize pain and distress of animals that undergo survival surgery. Local regulations can vary in the methods that must be used.

4   Notes

1. A small piece of Styrofoam can serve as a disposable surgical platform. However, it is important to assure that clean, sterile surgical surfaces are used. Sterile drapes are very useful for maintaining a sterile field.

2. Care must be used to avoid anesthetic overdose, which can be accomplished by precise control of the isoflurane concentration. It is important to follow general anesthesia procedures for laboratory mice during this surgical procedure and check the depth of anesthesia as described.

3. This procedure may induce respiratory distress so it must be performed both quickly and gently. Proper insertion of the angiocatheter into the trachea involves pointing the tip of the tube up as it enters the throat which will help to avoid insertion into the esophagus.

4. If the surgeon prefers that the angiocatheter is more rigid during intubation the angiocatheter can be prepared by initially removing the needle from the angiocatheter and then cut off the sharp bevel. The blunted needle can be reinserted into the angiocatheter before intubation and then remove the needle after intubation.

5. It is important to avoid approaching the sternal border too closely as the internal thoracic artery runs along the sternal border inside the thoracic cavity and can be easily damaged. Damaging the artery will induce extensive bleeding that can be difficult to control.

6. In the event that the superficial thoracic vein is damaged, cautery is usually the most effective way to restore hemostasis. Damaging this vein will produce bleeding; however this is not usually extensive or life-threatening and can be controlled using sterile cotton applicators.

7. The third intercostal space is the space between the third and fourth rib where the lowest part of the lung is observed.

8. When running the ligature under the LAD the insertion of the needle should be shallow since it is important to avoid entering the chamber of the left ventricle. At the same time, the ligation must not be too superficial as the ligature may cut through the wall of myocardium.

9. When tying the knot you should apply slight pressure to the thorax with the needle holder to push air out of the thorax to minimize the volume of air in the chest cavity and reduce the chance of pneumothorax.

References

1. Go AS, Mozaffarian D, Roger VL et al (2014) Heart disease and stroke statistics—2014 update a report from the American Heart Association. Circulation 129:28–292

2. Abarbanell AM, Herrmann JL, Weil BR et al (2010) Animal models of myocardial and vascular injury. J Surg Res 162:239–249

3. Smith FM (1918) The ligation of coronary arteries with electrocardiographic study. Arch Intern Med XXII:8–27. https://doi.org/10.1001/archinte.1918.00090120013002

4. Johns TNP, Olson BJ (1954) Experimental myocardial infarction: I. A method of coronary occlusion in small animals. Ann Surg 140:675–682

5. Maroko PR, Libby P, Ginks WR et al (1972) Coronary artery reperfusion. J Clin Invest 51:2710–2716

6. Pfeffer MA, Pfeffer JM, Fishbein MC et al (1979) Myocardial infarct size and ventricular function in rats. Circ Res 44:503–512

7. Michael LH, Entman ML, Hartley CJ et al (1995) Myocardial ischemia and reperfusion: a murine model. Am J Physiol 269:2147–2154

8. Degabriele NM, Griesenbach U, Sato K et al (2004) Critical appraisal of the mouse model of myocardial infarction. Exp Physiol 89:497–505

9. Kumar D, Hacker TA, Buck J et al (2005) Distinct mouse coronary anatomy and myocardial infarction consequent to ligation. Coron Artery Dis 16:41–44

10. Salto-Tellez M, Yung Lim S, El-Oakley RM et al (2004) Myocardial infarction in the C57BL/6J mouse: a quantifiable and highly reproducible experimental model. Cardiovasc Pathol 13:91–97

11. Xu Z, Alloush J, Beck E, Weisleder N (2014) A murine model of myocardial ischemia-reperfusion injury through ligation of the left anterior descending artery. J Vis Exp. https://doi.org/10.3791/51329

12. Virag JAI, Lust RM (2011) Coronary artery ligation and intramyocardial injection in a murine model of infarction. J Vis Exp. https://doi.org/10.3791/2581

13. Shao Y, Redfors B, Omerovic E (2013) Modified technique for coronary artery ligation in mice. J Vis Exp. https://doi.org/10.3791/3093

14. Tarnavski O, McMullen JR, Schinke M et al (2004) Mouse cardiac surgery: comprehensive techniques for the generation of mouse models of human diseases and their application for genomic studies. Physiol Genomics 16:349–360

15. Gao E, Lei YH, Shang X et al (2010) A novel and efficient model of coronary artery ligation and myocardial infarction in the mouse. Circ Res 107:1445–1453

16. Borst O, Ochmann C, Schönberger T et al (2011) Methods employed for induction and analysis of experimental myocardial infarction in mice. Cell Physiol Biochem 28:1–12. https://doi.org/10.1159/000331708

17. Diepenhorst GMP, van Gulik TM, Hack CE (2009) Complement-mediated ischemia-reperfusion injury: lessons learned from animal and clinical studies. Ann Surg 249:889–899

Chapter 13

A Rat Model of Perinatal Seizures Provoked by Global Hypoxia

Jason A. Justice and Russell M. Sanchez

Abstract

Hypoxic–ischemic encephalopathy (HIE) refers to acute brain injury that results from perinatal asphyxia. HIE is a major cause of neonatal seizures, and outcomes can range from apparent recovery to severe cognitive impairment, cerebral palsy, and epilepsy. Acute partial seizures frequently aid in indicating the severity and localization of brain injury. However, evidence also suggests that the occurrence of seizures further increases the likelihood of epilepsy in later life regardless of the severity of the initial injury. Here, we describe a neonatal rat model of seizure-provoking mild hypoxia without overt brain injury that has been used to investigate potential epileptogenic effects of hypoxia-associated seizures alone on neonatal brain development. Clinically, HIE is defined by brain injury, and thus, this model is not intended to mimic clinical HIE. Rather, its utility is in providing a model to understand the dynamic and long-term regulation of brain function and how this can be perturbed by early life seizures that are provoked by a commonly encountered pathophysiological trigger. Additionally, the model allows the study of brain pathophysiology without the potential confound of variable neuroanatomical changes that are reactive to widespread cell death.

Key words Hypoxia, Ischemia, Seizure, Epilepsy, Rat, Encephalopathy, Neonate

1 Introduction

Animal models of disease or disease pathophysiology have been invaluable to the advancement of medicine. The goals of specific models can be the artificial creation of a disease-mimicking state to study later stage processes or the induction of pathogenic processes to study how these generate the disease state. In this chapter, we describe a simple model in which transient global hypoxia in neonatal rat results in the occurrence of spontaneous seizures and modifies brain development in ways that could promote the pathogenesis of epilepsy without causing neuroanatomical injury.

Hypoxic–ischemic encephalopathy (HIE) refers to brain injury that occurs due to transient cerebral ischemia or hypoxemia during the perinatal period. HIE is a leading cause of neonatal seizures,

Binu Tharakan (ed.), *Traumatic and Ischemic Injury: Methods and Protocols*, Methods in Molecular Biology, vol. 1717,
https://doi.org/10.1007/978-1-4939-7526-6_13, © Springer Science+Business Media, LLC 2018

and can result in severe cognitive impairment, cerebral palsy, and epilepsy [1, 2]. Multiple animal models have been designed to mimic HIE to investigate injury mechanisms and the consequent pathology, or to test potential therapies [3]. For example, unilateral carotid artery ligation combined with global hypoxia in neonatal rat has been shown to provoke acute seizures and generate a reproducible infarct, and many of these animals develop epilepsy in later life [4]. The model described herein uses global hypoxia alone and does not produce an acute brain injury, yet also results in spontaneous seizures acutely and in later life [5]. Given the lack of anatomical injury, it is not intended to mimic clinical HIE. Rather its utility is in investigating the potential epileptogenic pathophysiological regulation of brain function in the absence of changes that are purely reactive to brain tissue loss.

Global hypoxia in postnatal day (P) 10 rat pups without experimental ischemia was reported by Jensen et al. [6] to provoke acute seizures and result in increased seizure propensity into adulthood. Notably, the acute ictogenic effect of hypoxia waned in pups at younger or older ages, mimicking an age-dependence observed clinically. Recently, continuous video-EEG monitoring has shown also that previously hypoxia-treated animals exhibited spontaneous recurrent seizures of increasing frequency within several weeks to 6 months after hypoxia [5]. Unlike those treated with ischemia by arterial occlusion, these animals became epileptic without evidence of neuroanatomical injury [7]. Furthermore, consequent modifications to hippocampal function can be observed in acute brain slices [8], allowing in vitro study of synaptic and network function. Thus, although this model does not precisely reproduce clinical HIE, it is often chosen to investigate how early life seizures with hypoxia can alter brain maturation to generate an epilepsy-prone state in the absence of anatomical injury.

2 Materials

1. Hypoxia Chamber (KE-25, Kent Scientific).

2. Oxygen Sensor.

3. Transducer (model TRN005, Kent Scientific).

4. Heating pad.

5. Video equipment (visual record of behavioral seizures).

6. EEG device (optional).

7. Rodent probe.

3   Methods

3.1   Preparation Before Experiment

Rat pups aged P10 are used as this age exhibits heightened seizure susceptibility (*see* **Notes 3** and **8**). Each animal is first weighed to ensure that they are appropriate weight for their age. Rectal temperature is measured using a standard small rodent probe (lubricated with glycerol) and thermometer before placing each animal in a chamber (*see* **Note 1**). Untreated animals should be handled identically to hypoxia-treated animals to control for potential effects of handling or maternal separation stress. Our approach is to use two chambers side by side, and to place a control animal in one chamber with the lid remaining off during the period of hypoxia treatment of the other animal. Both animals are returned to the dam together after hypoxia treatment.

3.2   Hypoxia Treatment

Animals are allowed to move freely within the chamber (or within a cage placed in the chamber). After closing the chamber lid, the valve on the nitrogen tank is opened to allow N_2 infusion to bring the O_2 concentration down to 7–9% within 20–30 s, and then gradually (so as not to overshoot) to 6–7% at which point a timer is started (*see* **Note 4**). The O_2 concentration is maintained at 6–7% for 4 min, and then lowered to 5–6% for 8 min. At this point (after 12 min total), the O_2 concentration is lowered by 1% per minute until the animal has become apneic for 30 s. Apnea is thought to be a reliable indicator of the magnitude of cerebral hypoxia, and thus, 30 s of apnea is used to ensure that each animal reaches the same depth of hypoxia. After 30 s of apnea, the chamber lid is opened to reintroduce the animal to room air (*see* **Note 2**).

3.3   Identification of Seizures

Rat pups will typically begin to exhibit behavioral signs of EEG abnormalities within 1–2 min. These begin with the appearance of myoclonic jerks that are characterized by sudden brief jerks of the whole body (*see* **Note 5**). Tonic–clonic seizures begin typically within 2–6 min and are correlated with electrographic seizures. These are characterized by the brief appearance of forelimb tonic extension followed by clonic side-to-side rapid head movement that lasts 10–60 s. This pattern will repeat throughout the duration of hypoxia prior to apnea with quite periods of no movement in between. An observer counts the numbers of tonic–clonic seizures as a measure of the severity of seizures. At the end of hypoxia treatment, rectal temperature is again measured (*see* **Note 6**). Each animal is earmarked for identification and returned to the dam (*see* **Note 7**).

4   Notes

1. Core temperatures that fall too far below physiological temperature can inhibit seizure activity. Conversely, hyperthermia can exacerbate seizures and even induce seizures without hypoxia in P10 rats [9].

2. The behavioral seizure phenotype has been shown clearly to indicate underlying electrographic seizures during hypoxia treatment. However, it is possible that subclinical EEG seizures could go undetected by relying solely on observations of behavior. For this reason, we exclude from analyses the small number of animals that did not exhibit tonic–clonic seizures.

3. It is possible that small differences in the optimal age exist between different rat strains, but this has not been studied. Similarly, possible gender differences have yet to be examined (we have used only males in our studies).

4. Often, animals will not exhibit seizures and may even become apneic in the first 1–2 min if the O_2 concentration is lowered too rapidly below 7%. The natural inclination is to increase hypoxia severity if no seizures are observed, but decreasing the rate of O_2 decrease is typically more effective when troubleshooting.

5. Individual myoclonic jerks are not always distinguishable from voluntary movements, as the animal may appear to jump. Only the tonic–clonic seizures (number and duration) are used to quantify the severity of seizures.

6. It is not uncommon for rectal temperature to drop between 1 and 2 degrees C during the hypoxia treatment. If it drops more than 2 degrees C, care should be taken to recalibrate the heating apparatus.

7. Seizures are also observed after return to room air, and can recur with decreasing frequency over 1–2 days.

8. Of note, this hypoxia protocol produces acute seizures in rats of different strains (with some potential variability in optimal age), but has not been observed to provoke seizures in mice. Our unpublished observations were that C57/Bl6 mouse pups exhibited recurrent forelimb tonus without clonic seizures during hypoxia treatment, and only exhibited repeated clonic seizures after return to room air. A modified protocol for mice that used alternating cycles from 5 to 7% hypoxia to 9% hypoxia over a 40-min period successfully provoked behavioral seizures [10] (It was not clear whether the seizures occurred during the lighter or deeper periods of hypoxia.).

Acknowledgment

This work was supported by resources and the use of facilities at the Central Texas Veterans Health Care System, Temple, TX, USA.

References

1. Tekgul H, Gauvreau K, Soul J et al (2006) The current etiologic profile and neurodevelopmental outcome of seizures in term newborn infants. Pediatrics 117:1270–1280

2. Ronen GM, Buckley D, Penney S et al (2007) Long-term prognosis in children with neonatal seizures: a population-based study. Neurology 69:1816–1822

3. Yager JY, Ashwal S (2009) Animal models of hypoxic-ischemic brain damage. Pediatr Neurol 40:156–167

4. Kadam SD, White AM, Staley KJ et al (2010) Continuous electroencephalographic monitoring with radio-telemetry in a rat model of perinatal hypoxia-ischemia reveals progressive post-stroke epilepsy. J Neurosci 30:404–415

5. Rakhade SN, Klein PM, Hyunh T et al (2011) Development of later life spontaneous seizures in a rodent model of hypoxia-induced neonatal seizures. Epilepsia 52:753–765

6. Jensen FE, Applegate CD, Holtzman D et al (1991) Epileptogenic effect of hypoxia in the immature rodent brain. Ann Neurol 29:629–637

7. Sanchez RM, Koh S, Rio C et al (2001) Decreased glutamate receptor 2 expression and enhanced epileptogenesis in immature rat hippocampus after perinatal hypoxia-induced seizures. J Neurosci 21:8154–8163

8. Jensen FE, Wang C, Stafstrom CE et al (1998) Acute and chronic increases in excitability in rat hippocampal slices after perinatal hypoxia in vivo. J Neurophysiol 79:73–81

9. Dube C, Richichi C, Bender RA et al (2006) Temporal lobe epilepsy after experimental prolonged febrile seizures: prospective analysis. Brain 129:911–922

10. Rakhade SN, Fitzgerald EF, Klein PM et al (2012) Glutamate receptor phosphorylation at serine 831 and 845 modulates seizure susceptibility and hippocampal hyperexcitability after early life seizures. J Neurosci 32:17800–17812

Experimental Protocol for Cecal Ligation and Puncture Model of Polymicrobial Sepsis and Assessment of Vascular Functions in Mice

Santosh Kumar Mishra and Soumen Choudhury

Abstract

Sepsis is the systemic inflammatory response syndrome that occurs during infection and is exacerbated by the inappropriate immune response encountered by the affected individual. Despite extensive research, sepsis in humans is one of the biggest challenges for clinicians. The high mortality rate in sepsis is primarily due to hypoperfusion-induced multiorgan dysfunctions, resulting from a marked decrease in peripheral resistance. Vascular dysfunctions are further aggravated by sepsis-induced impairment in myocardial contractility. Circulatory failure in sepsis is characterized by refractory hypotension and vascular hyporeactivity (vasoplegia) to clinically used vasoconstrictors. To investigate the complex pathophysiology of sepsis and its associated multiple organ dysfunction, several animal models have been developed. However, cecal ligation and puncture (CLP) model of murine sepsis is still considered as 'gold standard' in sepsis research. In this protocol we have described the standard surgical procedure to induce polymicrobial sepsis by cecal ligation and puncture. Further, we have described the protocol to study the molecular mechanisms underlying vascular dysfunctions in sepsis.

Key words Cecal ligation and puncture, Polymicrobial sepsis, Vascular reactivity

1 Introduction

Despite decades of research, treating sepsis and its associated multiple organ dysfunctions remains a challenge for the clinician at intensive care units. In USA, sepsis ranks tenth leading cause of death with mortality rates varying between 30% and 70% among ICU patients [1–3]. In Indian context, hospital mortality and 28 days mortality of severe sepsis were found to be 65.2% and 64.6%, respectively, within the period between June 2006 and June 2009 [4]. Since sepsis continues to be a substantial burden on healthcare, proper understanding of its complex pathophysiology needs to be explored to identify novel therapeutic targets and effective treatment strategy to combat this fatal condition.

Binu Tharakan (ed.), *Traumatic and Ischemic Injury: Methods and Protocols*, Methods in Molecular Biology, vol. 1717, https://doi.org/10.1007/978-1-4939-7526-6_14, © Springer Science+Business Media, LLC 2018

Due to the overlapping and multiple pathophysiology, use of suitable animal model which can reproduce the clinical conditions of sepsis has become an utmost importance in biomedical research. Several animal models replicate the clinical signs and laboratory findings commonly observed in animal and human sepsis. Such models include intravascular infusion of endotoxin [5] or live bacteria [6], bacterial peritonitis [7], cecal ligation and puncture [8], soft tissue infection [9], pneumonia model [10], and meningitis model [11]. In spite of evidences that endotoxin plays an important role in the pathogenesis of sepsis, several authors have expressed their concerns that the infusion of endotoxin is not a suitable model to study sepsis [12]. Cecal ligation and puncture (CLP) model is characterized by increased cardiac output and organ blood flow in early stage (i.e., hyperdynamic phase, 6 h after onset of sepsis [13]; and decreased tissue perfusion at the late stage (hypodynamic phase 18 h after onset of sepsis [14]. In this model, sepsis develops due to peritoneal contamination with mixed flora (anaerobic to facultatively to aerobic, gram-positive and gram-negative organisms) in the presence of devitalized or ischemic/necrotic tissue and thus bears an obvious resemblance to clinical reality. Furthermore, similar metabolic, immunological and apoptotic responses are observed in the CLP model as in human disease, strengthening the validity of this model [15–17]. In certain experimental conditions, researchers have modified the standard CLP procedure by conducting laparotomy after CLP and excising the necrotic cecum followed by peritoneal washing. Better outcome on survival time, under this condition, depends on the time period at which cecal resection is done; earlier the resection better is the survival [18, 19]. Another intervention is colon ascendens stent peritonitis (CASP) model where instead of puncturing cecum, a stent is introduced in the ascending colon distal to the ileocecal valve [20] to induce polymicrobial sepsis.

The inability of the circulatory system to provide adequate blood supply and hence oxygen and nutrient delivery to the tissues is considered as the main contributor to the progressive organ failure. The late, hypodynamic phase of sepsis is associated with the development of immunocompromised state in host [21]. Accumulating evidences suggest that decreased vascular responsiveness to vasopressors, a condition referred to as 'vasoplegia' [22] appears to be the alarming clinical sign in hypodynamic phase of sepsis. This vascular hyperreactivity appears to be directly or indirectly (via activation of the soluble guanylate cyclase enzyme) mediated by nitric oxide [23, 24], which is produced in large amounts during sepsis, mainly by the inducible isoform of the nitric oxide synthase (iNOS) enzyme. High level of G-protein receptor kinase-2 (GRK-2) expression in septic mouse aorta by nitric oxide to induce adrenoceptor desensitization is considered to be another possible mechanism contributing to cardiovascular hyporesponsiveness in septic shock [25]. However, the mechanism underlying this vascular hyporesponsiveness is yet to be fully explored. Here in this context, we have described the procedure to induce polymicrobial sepsis using cecal ligation and puncture model in mice and experimental protocols to assess the vascular dysfunctions due to sepsis.

2 Materials Required for Induction of Sepsis

1. Swiss albino mice.
2. Xylazine (10–15 mg/kg body weight; *i.p.*).
3. Ketamine (80–100 mg/kg body weight; *i.p.*).
4. Sterile saline solution (0.9% w/v).
5. 70% ethanol.
6. Razor blades or electric trimmer.
7. 21 G needles and black-braided silk suture (4-0).
8. Surgical instruments: scalpel, dissection scissor, straight surgical scissor, sterile surgical needle both straight and curved (22–27 G), needle holder.
9. Platform for animal surgery.
10. Water heated thermal blanket.
11. Syringes (1 mL), face mask, sterile gloves, and surgical gown.

3 Surgical Procedure

1. All the surgical maneuvers should be approved by the national and institutional ethical committee.
2. Anesthetize the mice with a combination of xylazine (10 mg/kg body weight, i.p.) and ketamine (80 mg/kg body weight, i.p.). With the availability of anesthetic chamber, inhalant anesthetic like isoflurane can be used as an alternative. Absence of reflex response to foot squeezing is an indication of adequate anesthesia.
3. Shave the lower half of the abdomen with the razor blades or electric trimmer without damaging underlining skin. Disinfect the shaved area of skin with 70% ethanol and place the animal on a platform for surgery (*see* Fig. 1a).
4. With the help of scalpel, make a midline incision (1.5–2 cm) over the skin without directly penetrating the peritoneal cavity. Using small scissor, extend the initial incision. Identify the midline white fascia (linea alba) of abdominal musculature and make another incision (~1 cm) over the abdominal muscle to get access to the peritoneal cavity (*see* Fig. 1b). Care must be taken not to damage the lower small bowel while giving incision over the abdominal muscle.

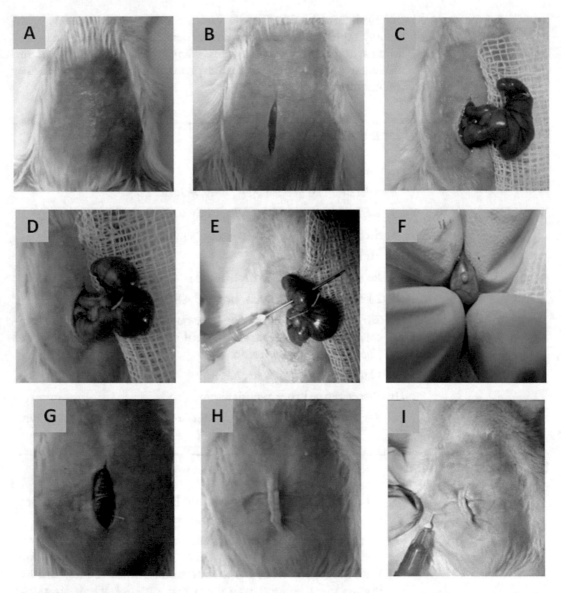

Fig. 1 Surgical procedures depicting induction of sepsis by cecal ligation and puncture. (**a**) Shaving and disinfecting the abdominal area (step 3). (**b**) Midline incision through skin (step 4). (**c**) Exteriorization of cecum (step 5). Note that after giving midline incision, cecum can be located mostly at left lower half of the abdominal cavity. (**d**) Ligation of the cecum (step 6). Note that site of the ligation is one of the major determinants of the severity of sepsis. (**e**) "Through and through" puncturing of cecum at the midway between ligation point and distal end of the cecum (step 7). We used to puncture the cecum twice with 21 G needle for induction of sepsis. (**f**) Extrusion of fecal material from the punctured hole to check the patency (step 7). Note that the amount of the extruded fecal material should be standardized to ensure consistency. (**g**) Closure of the peritoneum and abdominal muscle by simple running suture (step 9). (**h**) Closure of the skin by applying interrupted suture (step 9). (**i**) Subcutaneous injection of sterile saline (0.9% w/v)

5. With the help of a blunt scissors isolate the cecum and exteriorize it leaving the other parts of bowel (small and large) inside the peritoneal cavity. In most of the cases, the cecum is located at the left side of the abdomen (*see* Fig. 1c). Extensive care has to be taken not to damage cecal and mesenteric blood vessels to avoid severe bleeding complications.

6. Using 3-0 silk thread ligate the cecum at desired position depending upon the severity of sepsis induction as per the experimental design (*see* Fig. 1d). To prevent the obstruction of intestinal continuity, ligation should be placed below the ileocecal junction of mouse intestine.

   The location of ligation or length of the ligated cecum (i.e., distance between distal end of cecum and ligated point) determines the severity of sepsis in this model. A ratio of 50:50 between the ligated cecum and distance from ligation point to base of the cecum results in mid-grade (40% survival rate) sepsis while 75:25 ratio gives rise to high grade (100% mortality) sepsis [26].

   For the sham control procedure, replace the exteriorized cecum back into the abdominal cavity without doing ligation and puncture of the cecum. This set of animal is kept to ensure the proper aseptic surgery without causing undesired stress to the animals during the operating procedure as there should be no mortality in sham control mice.

7. Gently push the fecal material toward the distal part of cecum to uniformly distribute the cecal content and gently aspirate the trapped gas or air, if any. With the help of a sterile 21 G needle, puncture the cecum twice through and through at the midway between ligation and tip of the cecum (Fig. 1e). Care must be taken not to damage any blood vessels to avoid excess bleeding. Gently push a little amount of fecal materials through the hole at both the sides of puncture without damaging the wall of cecum. This is done to check the patency of the punctured hole (Fig. 1f). But remember that the amount of extruded cecal materials should be same for all the animals used during the experiment to ensure consistency.

8. Replace the ligated and punctured cecum back into the abdominal cavity. In case any fecal material is spread over the surgical wound over the skin, remove it with the help of sterile clean cloth or cotton swab.

9. Close the peritoneum and abdominal muscle layer by applying simple running suture while simple interrupted suture is used to close the skin wound (*see* Fig. 1g, h).

10. Resuscitate the operated mice by injecting subcutaneously prewarmed (37 °C) sterile 0.9% saline (1 mL/mouse). This helps to prevent hypovolemia from anesthesia, surgery and

subsequent shock (*see* Fig. 1i). Further, this fluid resuscitation induces hyperdynamic phase of sepsis [8, 27].

In some protocol, to avoid postoperative algesia, a subcutaneous injection of buprenorphine (0.05–0.2 mg/kg, every 6 h for at least 2 days) or tramadol (20 mg/kg) is given. But as these drugs can suppress respiration and locomotion, they may misinterpret the signs of sepsis [28, 29]. Further, to avoid the possible interaction of these drugs with test drugs, this step is omitted in the experimental protocol followed in our laboratory.

11. Gently place the mouse over its back on the thermal blanket and monitor until the animal recovers from anesthesia (regain of righting reflex). Usually the animal recovers from anesthesia within 30 min–1 h. Thereafter return the animal to a clean cage kept in a thermostatically controlled (22 °C) room under 12:12 h dark–light cycle. Provide free access to food and water and closely monitor the animal every half an hour for at least 2–3 h and thereafter at every 6 h.

3.1 Critical Points to Be Noted

Following points need to be kept in mind while performing cecal ligation and puncture model of sepsis as these have serious impact on the outcome of sepsis and the operating procedure.

1. The operating procedure for induction of sepsis should be carried out at same time point of the day as circadian cycle influences the outcome of inflammatory response [14].

2. Proper aseptic condition should be maintained during the operating procedure. The complete operating procedure for each animal can be performed within 10 min by an expert and experienced investigator.

3. It is to be noted that the ligation length, thickness or size of the needle, number of punctures determine the mortality rate [29–31].

4. The investigator needs to standardize the amount of released fecal material while squeezing the cecum after puncturing with the needle because it largely affects the mortality rate in CLP model of sepsis.

5. Excess manipulation of bowel during surgery should be avoided and proper care should be taken not to cause any unwanted bleeding which may influence the outcome of sepsis.

6. As mentioned earlier, postoperative fluid resuscitation to the operated animal should be done.

7. Substantial variability exists among different strains, sex, and age of the animal in respect to the outcome of sepsis and associated survival [32–35].

3.2 Assessment of Sepsis

The degree of sepsis following surgical procedure can be judged by evaluating several parameters. The most commonly used are as follows:

1. Survival time: The survival time following CLP varies with degree of sepsis.

2. Physical parameters and behavioral signs like body temperature, presence of conjunctivitis, absence of grooming activities with resulting ruffled fur, reduced intake of food and water and lethargic conditions.

3. Bacterial load in blood, peritoneal washes and organs of interest. Bacterial load is expressed as colony forming unit (CFU).

4. Cytokines and chemokines: Estimation of plasma cytokines like IL-1β, TNF-α, IL-10, and IL-6. Monocyte chemoattractant protein-1 (MCP-1) and high mobility group B-1 protein (HMGB1) are, among other, most commonly used parameters to evaluate the degree of sepsis.

5. Plasma and tissue nitrite level determination.

6. Estimation of serum lactate level: In septic patients and experimental animals, due to increased anaerobic metabolism there is enhanced deposition of lactate which lowers the pH value of blood/serum.

7. Determination of myeloperoxidase activity in the affected organs. It is a lysosomal protein stored in azurophilic granules of neutrophil. Increased activity of this enzyme is an indication of excess neutrophil infiltration which is considered to be a marker of inflammation.

4 Assessment of Vascular Reactivity in Sepsis

As mentioned earlier, sepsis is characterized by initial hyperdynamic phase followed by late hypodynamic phase. Reactivity of vascular beds to vasoactive substances thus varies accordingly among different stages of sepsis. Significant fall in the mean arterial blood pressure, impaired tissue perfusion and reduced reactivity of microcirculation to endogenous vasoconstrictor are some of the prominent signs of late stage of sepsis. Decreased vascular responsiveness to commonly used vasopressors (norepinephrine, vasopressin) in septic patient is considered to be one of the major causes of cardiovascular complications and circulatory failure. Nor-epinephrine acts via activation of *alpha*$_1$-adrenergic receptors in vascular beds to induce vasoconstriction. Among different subtypes (α_{1A}, α_{1B} and α_{1D}), α_{1D} adrenergic receptor subtype is most abundantly present [36, 37] and contributes to the vasoconstrictor response [38, 39] in mouse aorta. Thus dysfunction of this receptor subtype in vascular beds is considered to be one of the underlying mechanisms of vascular hyporeactivity in sepsis [40]. Here we describe the methods to be employed to assess the vascular reactivity in sepsis and to identify the molecular mechanisms responsible for vascular hyporeactivity.

4.1 Measurement of Blood Pressure to Determine Vascular Reactivity In Vivo

Blood pressure (BP) in laboratory animals can be recorded by three different methods viz. (a) tail cuff method (noninvasive), (b) intra-arterial catheters (invasive), and (c) radiotelemetry. Among these techniques, invasive blood pressure (IBP) monitoring is considered to be the gold standard method to evaluate the effect of different agonists and antagonists on blood pressure and to correlate with the vascular reactivity. The indirect or noninvasive blood pressure (NIBP) technique is most commonly used to screen the antihypertensive property of a drug given for a prolonged period. The major advantages of this technique are (1) it does not require surgery and can be used to obtain repeated measurement of systolic BP in conscious animals (2) it can be used among large numbers of animals to screen for systolic hypertension or substantial differences in systolic BP. However, as documented by Reddy and coworkers [41], tail cuff method is not suitable for measurement of diastolic BP. Further, the animals need to be trained well enough before recording, as parameters like BP and heart rate are significantly altered by the restrain stress [42]. On the other hand, direct method of blood pressure measurement (intra-arterial catheters or telemetry), though requires surgery, yields the most precise values and is able to quantify the relationship between BP and other variables (target organ damage, diet, genotype, etc.). This method has the advantage of constructing dose–response curves to vasoactive agonists such as norepinephrine, phenylephrine, and vasopressin that cause rise in blood pressure when administered intravenously. Although isoflurane is considered as preferable anesthetic for hemodynamic analysis in cardiac disorder model, pentobarbitone, among the injectable anesthetics, produces a less adverse effect on heart rate of mouse [43]. We used invasive technique for determining mean arterial pressure in septic animals for assessment of vascular responsiveness.

Protocol for Invasive Blood Pressure Recording in Rats

4.1.1 Materials Required

1. Swiss albino mice.

2. Drugs: heparin (5 IU/mL), norepinephrine (1 mg/mL), normal saline (0.9% w/v), pentobarbital sodium (60 mg/kg body weight; *i.p.*).

3. Surgical instruments: fine scissors, micro forceps, intravenous cannula made up of polypropylene tube (Cat no. SP0109; ADInstruments, Australia).

4. Powerlab physiograph with lab chart pro v7.3.7 software (ADInstruments, Australia), Pressure transducer (MLT844) with clip-on BP dome (ADInstruments, Australia) and sphygmomanometer.

5. Accessories: 1 mL tuberculin syringe, 10 mL syringe, 26 G needle with blunt end, surgical platform, dissecting stereomicroscope (Motic), three-way stopcock, razor blade, adhesive tape, surgical thread.

4.1.2 Calibration of Pressure Transducer

Remove the pressure cuff of the sphygmomanometer and connect the tubing to one end of the three-way stopcock fitted with the pressure transducer. Finally connect the pressure transducer to the Powerlab physiograph. Give a known pressure to raise the mercury level in the sphygmomanometer and record the change in the voltage (millivolts) shown by pressure transducer. Two point calibration between voltage and pressure thus can be set in the Lab Chart Pro software and then proceed for the following steps.

4.1.3 Cannulation of Vein

1. Anesthetize the mice with pentobarbital sodium (60 mg/kg, *i.p.*). Usually after 15–20 min the animal gets anesthetized. Check the absence of reflex action as mentioned earlier.

2. Place the animal in dorsoventral position and secure the fore and hind limbs.

3. For drug administration, either jugular or femoral vein is cannulated. However, femoral vein is preferred over jugular vein as carotid cannulation for determining the mean arterial pressure is sometimes inconvenient to perform if jugular vein is cannulated. Further, the right leg is preferred for femoral vain cannulation [44].

   For femoral vein cannulation, give an incision over the skin on right thigh and locate the femoral vein adjacent to femoral nerve. Put a nick toward head/heart end and introduce the vein catheter. Flush the cannula with normal saline to prevent thrombosis [45, 46].

   For jugular vein cannulation, shave the neck region and put an incision to locate the vein below the dermis. Isolate the vein from surrounding fat and tissue with the help of a blunt forceps. Tie a knot toward cephalic end while put a nick toward the heart and introduce the catheter and secure the catheter with the help of a thread [47].

4.1.4 Cannulation of Carotid Artery

1. Carefully shave the neck region and expose the underlying skin to separate the muscle by blunt incision. Care should be taken not to damage any underlying blood vessels to avoid bleeding.

2. Carotid artery (red) along with the vagus nerve (white) on either side of the trachea can be identified by observing the

pulsation of the artery. Separate the carotid artery from vagus nerve and adjacent connective tissue. Care should be taken not to stimulate the vagus nerve as its stimulation may decrease the heart rate and increases the risk of various respiratory abnormalities [48, 49].

3. Tie the cephalic end of carotid while toward the cardiac end put a knot to block the blood flow. It is to be noted that a bulldog clamp can also be used instead of knot, but as it may damage the artery, usually we avoid this practice. Put a small nick between the knots toward the heart with the help of a fine scissors.

4. Cannulate the carotid artery through the nick using a cannula prefilled with heparinized normal saline (0.5 IU/mL). Connect the other end of the cannula to a three-way stopcock/ saline filled tuberculin syringe.

5. Tie the cannula inserted in carotid artery with the help of saline presoaked thread without obstructing the blood flow in the carotid cannula. After cannulation, slowly release the bulldog clamp or loosen the knot at the cardiac end. Ensure that there is no bleeding at the cannulation site. Use some adhesive substance to fix the cannula and record the basal BP.

One can record ECG along with the BP. Three-lead bipolar ECG can be used to record the electrocardiograph. For this, place the positive, negative, and reference electrocardiogram electrodes at the left forearm, right forearm, and left thigh, respectively.

4.1.5 Recording Blood Pressure and Response of the Drug

1. Allow the whole setup to stabilize for 10–20 min. Record the baseline blood pressure for 10–15 min. With the help of online recording option of Lab Chart Pro software, measure the different parameters like systolic, diastolic pressure, heart rate, mean arterial pressure etc.

2. To record the dose response to nor-epinephrine (NE), administer different doses of NE (0.5–4 µg/kg/min) at a total volume of 0.1 mL each. After each dose of drug administration, inject 0.1 mL of normal saline through the venous cannula. Administer the next dose after an interval of 5 min and record the blood pressure responses to a maximum of four doses of the agonist per animal.

Usually septic animals exhibit lesser degree of response to NE compared to sham-operated animals.

4.2 In Vitro Method for Assessment of Vascular Reactivity

1. Swiss albino mice.

2. Xylazine (10–15 mg/kg body weight).

3. Ketamine (80–100 mg/kg body weight).

4.2.1 Materials Required

4. Modified Krebs-Henseleit solution (MKHS) with following composition (mmol/L): 118.0 NaCl, 4.7 KCl, 2.5

$CaCl_2 \cdot 2H_2O$, 1.2 $MgSO_4 \cdot 7H_2O$, 1.2 KH_2PO_4, 11.9 $NaHCO_3$, and 11.1 D-glucose; adjust the pH to 7.4.

5. Norepinephrine (NE), 1400 W, phenylephrine (PE), acetylcholine (ACh), sodium nitroprusside (SNP) from Sigma and methyl 5-[2-(5-nitro-2-furyl)vinyl]-2-furoat from Calbiochem.

6. Microscissors, Vannas scissors, dissecting forceps, and iris forceps.

7. "L"-shaped hooks made from 37 G stainless steel wire.

8. Dissection stereo microscope (Motic).

9. Powerlab polyphysiograph (ADInstruments, Australia); isometric force transducer (ADInstruments, Australia); Labchart Pro v7.3.7 and Graph pad prism 6 (La Jolla, CA, USA) software.

10. Two- or four-chambered thermostatically (37 ± 0.5 °C) controlled organ bath assembly (Ugo Basile, Italy).

11. Medical gas (74% N_2 + 21% O_2 + 5% CO_2) cylinder.

4.2.2 Calibration of Force Transducer

Before starting the experiments, two-point calibration of the force transducers need to be done as per the protocol prescribed by the manufacturer. Usually calibration is performed using a known weight (e.g., 2 g). Set the sampling rate to 5 s/sample, range to 20 mV with main filter on. After zeroing the all inputs, hang the known weight (1 or 2 g) to the transducer and record the change in the voltage. Perform the two-point calibration between the voltage (millivolts) and tension (g). Now the change in the tension will be displayed in 'g' unit.

4.2.3 Tissue Preparation and Mechanical Recording

1. Anesthetize the animal under xylazine–ketamine anesthesia. With the help of a scalpel, make a midline incision on the abdomen and extend the wound near the xyphoid region using a blunt scissor. Sacrifice the mice by bleeding from vena cava. Open the chest and collect the lung and heart *en bloc* along with thoracic aorta and immediately place in a Petri dish containing aerated ice cold (4 °C) MKHS solution.

2. Cut the thoracic aorta immediately below the aortic arch. Clean off the thoracic aorta from adhering tissues and fats under dissecting stereomicroscope using Vannas scissors and dissecting forceps. Carefully cut the thoracic aorta into four to five rings of 3–4 mm length. For the best results, the dissection procedure should be done in aerated ice cold MKHS solution. It is advisable to change the solution in the Petri dishes periodically during the isolation procedure. Fresh aortic rings can be used immediately within an hour of isolation or they can be stored in MKHS at 4 °C for 18–24 h before use.

3. Pass two "L"-shaped stainless steel hooks through the aortic rings in opposite direction under the microscope taking care not to damage the endothelium. Anchor one end of the tissue to the glass hook fitted to the thermostatically controlled (37 ± 0.5 °C) organ bath containing modified Krebs–Henseleit solution while the other end should be connected to the isometric force transducer with the help of a stainless steel wire. The solution should be continuously bubbled with medical gas (74% N_2 + 21% O_2 + 5% CO_2).

4. Apply a passive tension of 1.0 g to the tissue and allow the arterial rings to equilibrate under this resting tension for 60–80 min. During this equilibration period, the bath fluid needs to be changed once in every 15 min.

5. Measure the change in tension with the help of high-sensitivity isometric force transducer connected to Power lab and analyze the data using LabChart Pro V7.3.7 software program.

6. At the end of the equilibration period, adjust the basal tension to zero and check the tissue viability by eliciting contraction in the presence of high K^+ (80 mM) depolarizing solution (*see below*). A contraction between 0.5 and 1.0 g is considered to be ideal for further experiments. However, it is to be noted that aortic rings from CLP mice at hypodynamic stage show lesser contraction (0.2–0.3 g) in the presence of high K^+ solution. On attaining contraction-plateau, replace the high K^+-solution by normal MKHS and wash the arterial rings for several times with normal MKHS to restore baseline resting tension.

   High K^+ (80 mM) solution may be prepared by replacing equimolar NaCl by KCl. The composition (mmol/L) is as follows: 43.9 NaCl, 78.8 KCl, 2.5 $CaCl_2 \cdot 2H_2O$, 1.2 $MgSO_4 \cdot 7H_2O$, 1.2 KH_2PO_4, 11.9 $NaHCO_3$, and 11.1 glucose; adjust the pH to 7.4.

4.2.4 Assessment of Vascular Reactivity

1. To assess the vascular reactivity, record the concentration-dependent (0.1 nM–1 μM) contractile response to norepinephrine (NE) in aortic rings from sham operated (SO) and septic (CLP) mice. For this, expose the arterial rings to cumulative concentration of NE ranging from 0.1 nM to 10 μM at an increment of 0.5 log unit. Usually NE produces maximum contraction (E_{max}) at 1 μM concentration, thereafter subsequent addition of higher concentration exerts decline in the contractile response. Several reports as well as findings from our laboratory have documented that when tissue samples are collected at hypodynamic phase of sepsis, arterial rings shows reduction in maximum contractile response (E_{max}) compared to SO mice.

For the assessment of involvement of iNOS-derived nitric oxide (NO) in mediating such vascular hyporeactivity in sepsis, arterial rings may be preincubated with 1400 W (a specific iNOS inhibitor) for 30 min before recording the cumulative concentration-response to NE. We have reported that in vitro preincubation of 1400 W (10 μM) dramatically restored the vascular hyporeactivity to NE in arterial rings from CLP mice without affecting the response in SO mice.

Similarly, to study the role of receptor desensitization mechanism in impairing NE-induced contraction in sepsis, preincubate the arterial rings with methyl 5-[2-(5-nitro-2-furyl)vinyl]-2-furoate (a selective inhibitor of GRK-2) for 30 min before recording the concentration-dependent contractile response to NE. Recently, we have documented that preincubation of arterial rings with GRK-2 inhibitor (1 μM) partially restored the vascular hyporeactivity to NE in septic mice [50, 51].

We have reported that this observed vascular hyporeactivity to NE is well correlated with the down regulation of α_{1D}-adrenoceptors in mouse aorta from septic mice [40]. This can be assessed by measuring the relative expression of α_{1D} mRNA in mouse aorta using real time RT-PCR. Using immunohistochemical technique, recently we have provided the evidence for decreased level of receptor protein (α_{1D} adrenoceptor) expression in mouse aorta from septic mice which is considered to be one of the underlying mechanisms for vascular hyporeactivity [50].

4.2.5 Assessment of Endothelial Dysfunction

Acetylcholine (ACh) is most commonly used to study the endothelial integrity of the blood vessels under investigation. By acting on endothelial muscarinic (M_3) receptors, it enhances intracellular Ca^{2+} concentration, and the Ca-calmodulin complex subsequently stimulates endothelial nitric oxide synthase (eNOS) activity to cause synthesis of NO. Nitric oxide, thus released from endothelium, causes vascular smooth muscle relaxation. Protocol for assessment of endothelial integrity/function is as follows;

1. After equilibration and assuring the tissue viability, precontract the tissue with phenylephrine. On reaching the steady state contraction, add ACh (1 nM–10 μM) cumulatively at an increment of 0.5 log unit. Aortic rings from SO mice usually relax between 90% and 100% on exposure to ACh. However CLP mice show merely 50–55% relaxation.

   It is to be noted that while comparing the vasodilator responses of an agonist between sham and septic animals, matching precontraction level with the contractile agent (e.g., in this case phenylephrine) should be obtained. Thus while recording the vasorelaxant response to either ACh or SNP, different concentration of phenylephrine (1 and 3 μM) need to be used to attain matching precontraction level in SO and

CLP groups, respectively. Different concentrations of phenylephrine are used because sepsis reduces the maximal response to the vasoconstrictor agents.

Another point to be noted while doing the experiments to study the vasodilator response is that agonist-induced precontraction should not display slow time-dependent decay. An ideal concentration–response curve to the vasodilator should be sigmoid in nature. If it is a linear response, discard the tissue. Thus it is always advisable to run a time control set while recording the vasorelaxant response to an agonist.

2. In order to ascertain that impairment in the relaxation response to ACh mediated by NO is due to endothelial dysfunction in sepsis, evaluate the vasodilator potency of sodium nitroprusside (SNP) which is an endothelium-independent nitrovasodilator. It relaxes vascular smooth muscle through stimulation of sGC with a consequential increase in tissue cGMP levels. In order to study the effect of sepsis on NO/cGMP signaling at the vascular cell level, record the concentration-dependent relaxations to SNP (0.01 nM–0.1 µM) in arterial rings precontracted with phenylephrine as mentioned earlier.

Reduced mRNA expression of eNOS and enhanced expression of iNOS mRNA provides further evidence of endothelial dysfunction in sepsis. Using commercially available kits or biochemical methods [50], one can estimate eNOS and iNOS-derived nitric oxide in the form of stable product nitrate to further strengthen the hypothesis.

4.3 Immuno-histochemistry for Alpha$_{1D}$-Adrenoceptor in Mouse Aorta

4.3.1 Materials Required

1. 3-amino propyl triethoxysilane (APES) precoated slide (Silane Prep Slides, Sigma), Coplin jar, microoven.

2. Xylene, ethanol, citric acid, sodium citrate, formalin, hydrogen peroxide, methanol, phosphate buffer saline (PBS).

3. DAB kit, Primary and secondary antibodies, Mayer's hematoxylin stain.

4. Microscope, microtome.

4.3.2 Experimental Protocol

1. Anesthetize the animal by xylazine–ketamine anesthesia. After proper anesthesia, isolate the thoracic aorta as described previously (*see protocol for in vitro vascular reactivity*). For immunohistochemistry experiment, aortic rings should be preserved in buffered formalin (NaH_2PO_4 4 g; Na_2HPO_4 6.5 g; formaldehyde (37%) 100 mL; distilled water to 900 mL; adjust the pH to 6.8–7.0). After proper fixation (after around 1 week), prepare the transverse tissue section (5 µm thick) on 3-amino propyl triethoxysilane (APES) precoated slide (Silane Prep Slides, Sigma). It is advisable to change buffered formalin once in a week during fixation.

2. Deparaffinize the tissue sections by keeping the slides in a coplin jar containing xylene for 5 min. Repeat the process twice. The slides are then rehydrated in descending grade of ethanol (100–70%) each for 5 min, followed by washing in deionized water for 5 min.

3. In the next step antigen retrieval is done by keeping the slides in antigen retrieval solution (*see below*) and boiled for 5 min. Care should be taken that during this boiling process tissue section should not slough off from the slides. Repeat the process thrice and allow the slides to cool at room temperature for 30 min. Wash the slides with deionized water three times for 5 min each.

   Antigen retrieval solution can be prepared by mixing 18 mL of citric acid (100 mM) and 82 mL of sodium citrate (100 mM) to a final volume of 1 L. Adjust the volume with deionized water. Adjust the pH to 6.0.

4. Arrange the slides in a humid chamber and 3% H_2O_2 in methanol is poured onto it. Allow equilibration for 20–30 min. Slides are then washed in 0.1 M PBS (pH—7.4) three times for 3–5 min each. Wipe the slides with tissue paper to soak excess water.

5. The sections are then encircled with hydrophobic pen. For protein blocking pour 5% goat serum protein on the slides and keep for 1 h.

6. Discard the blocking solution and wash the slides with PBS for three times. Incubate the tissue section with rabbit anti-mouse polyclonal antibody against α_{1D}-AR (1:50, Abcam, USA) for overnight at 4 °C in a moist chamber. Sections are then washed with 0.1 M PBS (pH 7.4) three times for 3–5 min each.

7. Slides are then washed three times with PBS and sections are incubated with secondary antibody tagged with HRP (goat anti-rabbit polyclonal IgG, Enzo Life Sciences, USA) for 30 min at room temperature in moist chamber. Wash the tissue sections with 0.1 M PBS (pH 7.4) three times for 3–5 min each.

8. After wiping the slides, pour ImmPACT DAB (1 drop in 1 mL of DAB buffer) (ImmPACT DAB, Peroxidase Substrate Kit, Vector Lab, USA) and 2.5 µL of H_2O_2 (in 1 mL of buffer) onto the slides and keep for 20–30 s. Stop the reaction by keeping the slides in running water. Slides are then washed with deionized water for 5 min.

9. Counter stain the section with Mayer's hematoxylin stain for 30–40 s. Rinse the slides with deionized water. Sections are

then dehydrated in ascending grade of alcohol (50–100%) for 3–5 min each.

10. Clean the slides in xylene for 5 min with two subsequent washings and mount in DPX. For negative control, slides are to be treated similarly with normal goat serum. Examine the tissue sections for protein signals under light microscope (OLYMPUS, IX51, USA). The density of α_{1D}-AR protein can be determined at five different areas located in four sections from each aorta under 20× magnification. The receptor density in different groups can be measured semiquantitatively by using ImageJ software (ImageJ, NIH, USA).

4.4 Reverse Transcriptase Real-Time PCR (RT-qPCR) Analysis for Detecting Vascular Dysfunction

Experimental Protocol

4.4.1 RNA Isolation

1. Isolate the thoracic aorta from mice of different groups in 0.1% diethyl pyrocarbonate-treated autoclaved PBS. Remove the adjacent adipose tissues under microscope and store the samples in RNAlater™ (RNA stabilization solution, Sigma, USA) at −80 °C or follow the subsequent steps for isolation of RNA.

   It is to be noted that while isolating RNA from tissue samples, proper aseptic measures should be practiced in the laboratory as RNA is comparatively more unstable than DNA. Hence, eppendorf tubes, tips and all other plasticware to be used during RNA isolation should be made RNAase free by dipping them in diethyl pyrocarbonate (DEPC, 0.1%) for overnight and then dry and autoclave them before use. As DEPC is carcinogenic always use sterilized gloves to avoid its contact with skin. It is advisable to use face mask to avoid contamination from saliva as it also contains RNAse enzyme. At the final step of RNA isolation (i.e., while adding DNAase- and RNAase-free water), use fresh gloves.

2. Several methods for RNA isolation from animal tissues exist. However, in our laboratory we are isolating the RNA samples from mouse aorta using commercially available kits. Mini RNA isolation kit manufactured by IBI Scientific (USA) is an example of such kit; however, such kits are also available from many other manufacturers (Ambion, Invitrogen, Qiagen, Zymo Research, etc.). Follow the instructions provided by the manufacturer to isolate the RNA sample. It is always advisable to treat RNA samples with RNase-free DNase to avoid contamination with genomic DNA.

3. Check the purity of the RNA by taking the absorbance at 260 and 280 nm and only samples showing A_{260}/A_{280} ratio 1.95–2.05 and A_{260}/A_{230} ratio >1.5 should be used for cDNA synthesis. One can quantify the RNA in the sample using the formula: $1OD_{260} = 40$ μg/mL.

4.4.2 cDNA Synthesis

We use commercially available kit (Revertaid® First strand cDNA synthesis kit, Thermo Scientific, USA) for synthesis of cDNA in our lab. It uses Moloney murine leukemia virus reverse transcriptase enzyme for amplification of cDNA from RNA. Use the manufacturer's guideline to add the reagents and equal quantity of RNA for synthesis of cDNA from different samples.

4.4.3 Reverse Transcriptase Real-Time PCR (RT-qPCR)

The real-time analysis of gene of interest can be performed either by doing absolute or relative quantification. We employed relative quantification of $alpha_{1D}$ adrenoceptor, GRK-2, eNOS, and iNOS gene expression to assess the molecular mechanisms of altered vascular reactivity in sepsis using SYBR Green chemistry. Relative expression of gene of interest can be quantified in relation to the expression of a housekeeping gene like GAPDH or β-actin etc. Each sample should be run in triplicate. Prepare PCR master mix by adding 12.5 μl SYBR Green/ROX master mix (Thermo Scientific, USA), 1.0 μL from 10 pmol/μL stock solution of each of the gene-specific forward and reverse primers, and 1 μL of cDNA (<500 ng/reaction) and adjust the final volume to 25 μL with RNAse-free water (Table 1). Prepare a nontemplate control (NTC) which contains all the ingredients of reaction mixture as stated above except template or cDNA.

Depending on the annealing temperature of the specific primer pairs, it is always advisable to validate or standardize the primer pairs and annealing temperature before running real-time PCR. This can be done by running simple gradient PCR and observing the specific band by running the PCR product in agarose gel electrophoresis.

One can design the primer sequence of his own using the gene information available in NCBI. However, we used the following primer sequence for analysis of the following genes in mouse aorta [50]:

Table 1
Details of the primer pairs for RT-qPCR

Type of gene	Primer sequences	Amplicon size	Accession No.
GAPDH	F 5′-AACTTTGGCATTGTGGAAGG-3′ R 5′ACACATTGGGGGTAGGAACA 3′	223 bp	GU214026.1
α_{1D} AR	F 5′-GCCTCTGAGGTGGTTCTGAG-3′ R 5′-GGACGAAGAAAAAGGGGAAC-3′	208 bp	AB030642.2
GRK2	F5′-GGCGATACTTCTACTTGTTCCC-3′ R 5′-CGTTCCTTGATCTGTGTCTCTT-3′	118 bp	NM130863.1
eNOS	F 5′-GGCTGGGTTTAGGGCTGT-3′ R 5′-GCTGTGGTCTGGTGCTGGT-3′	107 bp	NM008713.4
iNOS	F 5′-ACATCGACCCGTCCACAGTAT-3′ R 5′-CAGAGGGGTAGGCTTGTCTC-3′	177 bp	NM010927.3

Table 2
RT-qPCR reaction condition

Segment 1	Segment 2	Segment 3
95 °C/10 min. 1 cycle	95 °C: 15–30 s 55–60 °C*: 30 s. 72 °C: 20–30 s 40 cycles	95 °C: 1 min 55–65 °C: 30 s 95 °C: 30 s 1 cycle

1. Set up the program as per the manufacturer's instructions. A typical program is as follows in Table 2:

2. Run the PCR reaction and record the C_T value for each sample.

3. Calculate the relative expression using method described by Pfaffl [52] or Livak [53].

4.5 Representative Results

Here we describe the effect of sepsis on different parameters. Note that these results are simply a representative of these parameters. These may vary among different laboratories and experimental set up. We have performed CLP in Swiss albino mice. Cecum was punctured twice through and through with 21 G needle at the midway between ligated point and distal end of the cecum. As depicted in Fig. 2, we have observed that the mean survival time in septic mice (CLP) was 20.00 ± 1.66 h ($n = 22$) while all the sham-operated (SO) mice survived during the 72 h of observation period. Further we have observed that septic mice had significantly higher level of serum lactate (4.64 ± 0.05 mmol/L, $n = 6$ vs. sham (1.61 ± 0.37 mmol/L, $n = 6$), and blood bacterial count (6.91 ± 0.23 log CFU/mL; $n = 8$ vs. sham (no bacterial growth was observed) than the SO mice [50].

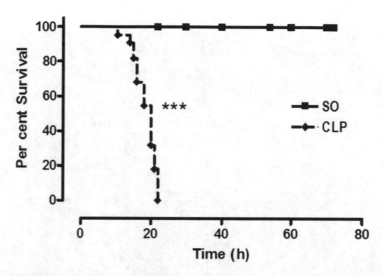

Fig. 2 Effect CLP-induced sepsis on survival time in mice. Sepsis was induced by puncturing cecum twice "through and through" with a 21-gauge needle. The overall difference in survival rate was determined by the Kaplan–Meier test followed by the log-rank test

It is well documented that cardiovascular complicacy is one of the leading causes of death, and is often associated with vascular hyporesponsiveness to commonly used vasopressor agents in septic patients. In a recently published article [50], we sought to describe the aortic vascular reactivity to norepinephrine (NE) and its underlying mechanisms in mouse subjected to CLP. As vascular hyporesponsiveness is typically seen in late phase of CLP model of sepsis, all the samples were collected at 20 h post-CLP. The mean arterial pressure of septic mouse was recorded as (41 ± 4.47 mmHg, $n = 7$) in comparison with SO controls (87.25 ± 5.45 mmHg, $n = 4$) [40]. In vitro exposure of the isolated aortic rings to nor-epinephrine (NE) revealed that maximum contraction to NE was significantly reduced in septic mice (*see* Fig. 3). On the other hand, endothelial dysfunction was prominent in septic mice as evidenced by decline in the acetylcholine (ACh)-induced relaxation in mouse aorta (*see* Fig. 4). However, there was no change in sodium-nitroprusside (SNP)-induced relaxation (*see* Fig. 5), which is an endothelium-independent vasodilator. Aortic hyporeactivity was further confirmed by the reduced protein (*see* Fig. 6) and mRNA (*see* Fig. 7a) expressions of $alpha_{1D}$-adrenoceptor in mouse aorta. Moreover, we have also found a significantly ($p < 0.001$) higher plasma norepinephrine level (144.90 ± 12.76 pg/mL vs 80.52 ± 1.71 pg/mL; $n = 6$) in septic mice with corresponding increase in GRK-2 mRNA expression in septic mouse aorta (*see* Fig. 7b). This is responsible for α_{1D}-adrenoceptor desensitization resulting in vascular hyporesponsiveness (*see* Fig. 7b). Aortic segments from CLP mice also showed reduced level of eNOS and higher level iNOS mRNA expression (*see* Fig. 7c, d).

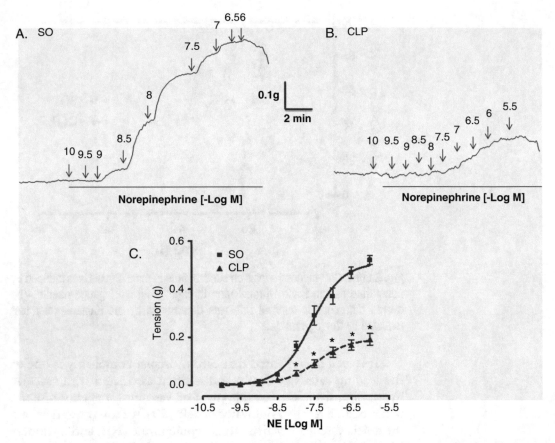

Fig. 3 Effect of sepsis on reactivity of aortic smooth muscle to norepinephrine (NE) in mice. Sepsis was induced by cecal ligation and puncture. Representative tracings show the concentration-dependent contractions to NE (0.1 nM–1 μM) in the aorta from SO (**A**) and CLP (**B**) mice. The line diagram (**C**) depicts the mean concentration–response curves elicited with cumulatively added NE on the aortic rings obtained from SO and CLP mice. Vertical bars represent SEM. Data were analyzed by two-way ANOVA followed by Bonferroni post hoc tests. *$p < 0.05$ in comparison to SO

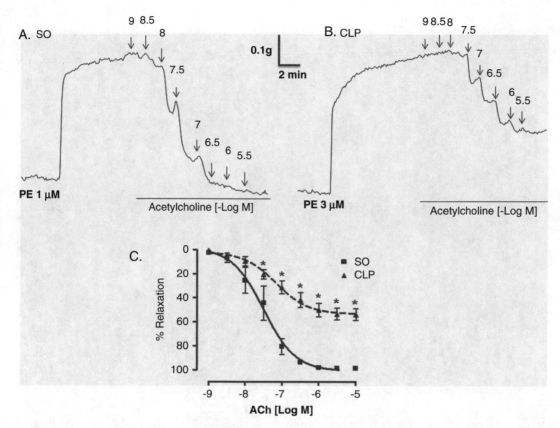

Fig. 4 Effect of sepsis on ACh-induced endothelial-dependent relaxation in mouse aortic rings. Sepsis was induced by cecal ligation and puncture. Representative tracings showing the concentration-dependent relaxations to cumulatively added ACh (0.1 nM–10 μM) in the phenylephrine (PE) pre-contracted aorta taken from SO (**A**) and CLP (**B**) mice. The line diagram (**C**) depicts the mean concentration–response curves of ACh on the aortic rings obtained from different groups. Vertical bars represent SEM. Data were analyzed by two-way ANOVA followed by Bonferroni post hoc tests. *$p < 0.05$ in comparison to SO

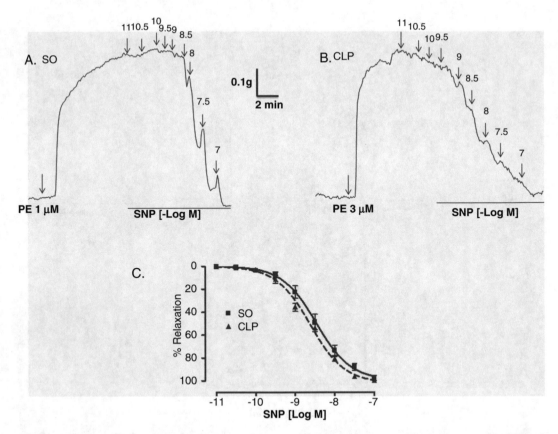

Fig. 5 Effect of sepsis on SNP-induced relaxations in mouse aortic rings pre-contracted with phenyleph-rine (PE). Raw tracings show the concentration-dependent relaxations to cumulatively added SNP (0.01 nM–0.1 μM) in the aorta taken from SO (**A**) and CLP (**B**) mice. The line diagram (**C**) depicts the mean concentration–response curves of SNP on the aortic rings obtained from different groups. Vertical bars represent SEM. Data were analyzed by two-way ANOVA followed by Bonferroni post hoc tests. Note that SNP-induced relaxation does not differ significantly between these groups

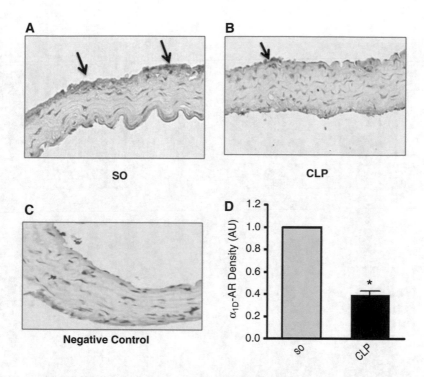

Fig. 6 Effect of sepsis on *alpha*$_{1D}$ adrenoceptor protein expression in mouse aorta. Representative immuno-histochemical images demonstrate the *alpha*$_{1D}$ receptor protein expression in aortic sections from SO (**A**) and CLP (**B**) mice. (**C**) represents negative control where the primary antibody was replaced by normal IgG. Arrow indicates the positive staining for alpha1D adrenoceptor in vascular smooth muscle. The histogram shows mean density of the *alpha*$_{1D}$ receptor protein (**D**). Note that sepsis reduced the density of α_{1D}-adrenoceptor protein as depicted in the bar diagram. Data were analyzed by one-way ANOVA followed by Newman–Keuls multiple comparison test. Vertical bars represent SEM. *$p < 0.001$ compared to SO

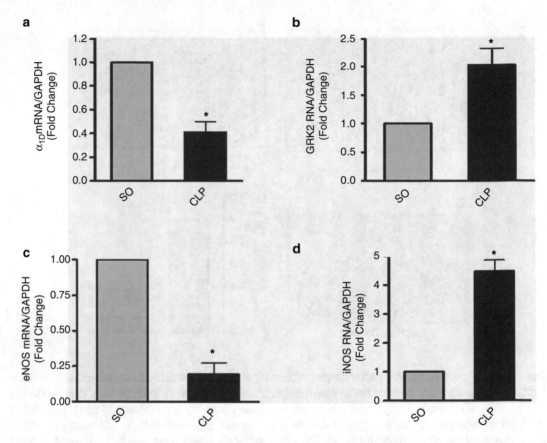

Fig. 7 Bar diagrams showing the effect of sepsis on relative expression of mRNA of *alpha* 1D (**a**), GRK-2 (**b**), eNOS (**c**), and iNOS (**d**) in mouse aorta. Data were analyzed by Student's *t*-test. *$p < 0.05$ in comparison to SO

References

1. Angus DC, Wax RS (2001) Epidemiology of sepsis: an update. Crit Care Med 29: S109–S116

2. Kochanek KD, Smith BL (2004) Deaths: preliminary data for 2002. Natl Vital Stat Rep 52:1–47

3. Finfer S, Bellomo R, Lipman J, French C, Dobb G, Myburgh J (2004) Adult population incidence of severe sepsis in Australian and New Zealand intensive care units. Intensive Care Med 30:589–596

4. Todi S, Chatterjee S, Sahu S, Bhattacharyya M (2010) Epidemiology of severe sepsis in India: an update. Crit Care 14:382

5. Taylor FB (2001) Staging of the pathophysiologic responses of the primate microvasculature to *Escherichia coli* and endotoxin: examination of the elements of the compensated response and their links to the corresponding uncompensated lethal variants. Crit Care Med 29: S78–S89

6. Silverstein R, Wood JG, Xue Q, Norimatsu M, Horn D, Morrison DC (2000) Differential host inflammatory responses to viable versus antibiotic-killed bacteria in experimental microbial sepsis. Infect Immun 68:2301–2308

7. Lang CH, Bagby GJ, Bornside GH, Vial LJ, Spitzer JJ (1983) Sustained hypermetabolic sepsis in rats: characterization of the model. J Surg Res 35:201–210

8. Wicherman KA, Baue AE, Chaudry IH (1980) Sepsis and septic shock-A review of laboratory models and a proposal. J Surg Res 29:189–201

9. Durkot MJ, Wolfe RR (1989) Hyper and hypodynamic models of sepsis in guinea pigs. J Surg Res 46:118–122

10. Karzai W, Cui X, Mehlhom B, Straube E, Hartung T, Gerstenberger E, Banks SM, Natanson C, Reinhart K, Eichacker PQ (2003) Protection with antibody to tumor necrosis factor differs with similarly lethal *Escherichia coli* versus *Staphylococcus aureus* pneumonia in rats. Anesthesiology 9:81–89

11. Ribes S, Domenech A, Cabellos C, Tubau F, Linares J, Viladrich PF, Gudiol F (2003) Experimental meningitis due to a high-level cephalosporin-resistant strain of *Streptococcus pneumoniae* serotype 23F. Enferm Infecc Microbiol Clin 21:329–333

12. Fink MP, Heard SO (1990) Laboratory models of sepsis and septic shock. J Surg Res 49: 186–196

13. Wang P, Ba ZF, Chaudry IH (1995) Endothelium-dependent relaxation is depressed at the macro- and microcirculatory levels during sepsis. Am J Physiol 269:R988–R994

14. Hubbard WJ, Choudhry M, Schwacha MG, Kerby JD, Rue LW, Bland KI, Chaudry IH (2005) Cecal ligation and puncture. Shock 24:52–57

15. Ayala A, Chaudry IH (1996) Immune dysfunction in murine polymicrobial sepsis: mediators, macrophages, lymphocytes and apoptosis. Shock 6:S27–S38

16. Hotchkiss RS, Karl IE (2003) The pathophysiology and treatment of sepsis. N Engl J Med 348:138–150

17. Buras JA, Holzmann B, Sitkovsky M (2005) Animal models of sepsis: setting the stage. Nat Rev Drug Discov 4:854–865

18. Xiao H, Siddiqui J, Remick DG (2006) Mechanisms of mortality in early and late sepsis. Infect Immun 74:5227–5235

19. Latifi SQ, O'Riordan MA, Levine AD (2002) Interleukin 10 controls the onset of irreversible septic shock. Infect Immun 70:4441–4446

20. Zantl N, Uebe A, Neumann B, Wagner H, Siewert J-R, Holzmann B, Heidecke C-D, Pfeffer K (1998) Essential role of gamma interferon in survival of colon ascendens stent peritonitis, a novel murine model of abdominal sepsis. Infect Immun 66:2300–2309

21. Shelley O, Murphy T, Paterson H, Mannick JA, Lederer JA (2003) Interaction between the innate and adaptive immune systems is required to survive sepsis and control inflammation after injury. Shock 20:123–129

22. Cobb JP, Danner RL (1996) Nitric oxide and septic shock. JAMA 275:1992–1996

23. da Silva-Santos JE, Terluk MR, Assreuy J (2002) Differential involvement of guanylate cyclase and potassium channels in nitric oxide-induced hyporesponsiveness to phenylephrine in endotoxemic rats. Shock 17:70–76

24. Chen SJ, Chen KH, CC W (2005) Nitric oxide-cyclic GMP contributes to abnormal activation of Na^+-K^+-ATPase in the aorta from rats with endotoxic shock. Shock 23:179–185

25. Secco DD, Olivon V, Correa T, Celes MR, Abreu M, Rossi M, Oliveira AM, Cunha F, Assreuy J (2010) Cardiovascular hyporesponsiveness in sepsis is associated with G-protein receptor kinase expression via a nitric oxide-dependent mechanism. Crit Care 14:P34

26. Rittirsch D, Huber-Lang MS, Flierl MA, Ward PA (2009) Immunodesign of experimental sepsis by cecal ligation and puncture. Nat Protoc 4:31–36

27. Hollenberg S, Dumasius A, Easington C, Colilla S, Neumann A, Parrillo J (2001) Characterization of a hyperdynamic murine model of resuscitated sepsis using echocardiography. Am J Respir Crit Care Med 32:2589–2597

28. Hugunin KM, Fry C, Shuster K, Nemzek JA (2010) Effects of tramadol and buprenorphine on select immunologic factors in a cecal ligation and puncture model. Shock 34:250–260

29. Toscano MG, Ganea D, Gamero AM (2011) Cecal ligation puncture procedure. J Vis Exp 51:e2860. https://doi.org/10.3791/2860.

30. Baker CC, Chaudry IH, Gaines HO, Baue AE (1983) Evaluation of factors affecting mortality rate after sepsis in a murine cecal ligation and puncture model. Surgery 94:331–335

31. Cuenca AG, Delano MJ, Kelly-Scumpia KM, Moldawer LL, Efron PA (2010) Current protocols in immunology: cecal ligation and puncture. Curr Protoc Immunol. https://doi.org/10.1002/0471142735.im1913s91

32. Diodato MD, Knoferl MW, Schwacha MG, Bland KI, Chaudry IH (2001) Gender differences in the inflammatory response and survival following haemorrhage and subsequent sepsis. Cytokine 14:162–169

33. Turnbull IR, Wlzorek JJ, Osborne D, Hotchkiss RS, Coopersmith CM, Buchman TG (2003) Effects of age on mortality and antibiotic efficacy in cecal ligation and puncture. Shock 19:310–313

34. Watanabe H, Numata K, Ito T, Takagi K, Matsukawa A (2004) Innate immune response in Th1- and Th2-dominant mouse strains. Shock 22:460–466

35. De Maio A, Torres MB, Reeves RH (2005) Genetic determinants influencing the response to injury, inflammation, and sepsis. Shock 23:11–17

36. Hosoda C, Tanoue A, Shibano M, Tanaka Y, Hiroyama M, Koshimizu TA, Cotecchia S, Kitamura T, Tsujimoto G, Koike K (2005) Correlation between vasoconstrictor roles and mRNA expression of α1-adrenoceptor subtypes in blood vessels of genetically engineered mice. Br J Pharmacol 146:456–466

37. Yamamoto Y, Koike K (2001) Characterization of α_1-adrenoceptor-mediated contraction in the mouse thoracic aorta. Eur J Pharmacol 424:131–140

38. Tanoue A, Nasa Y, Koshimizu T, Shinoura H, Oshikawa S, Kawai T, Sunada S, Takeo S, Tsujimoto G (2002) The α_{1D}-adrenergic receptor directly regulates arterial blood pressure via vasoconstriction. J Clin Invest 109:765–775

39. Cohn HI, Harris DM, Pesant S, Pfeiffer M, Rui-Hai Z, Koch WJ, Dorn GW, Eckhart AD (2008) Inhibition of vascular smooth muscle G protein-coupled receptor kinase 2 enhances α_{1D}-adrenergic receptor constriction. Am J Physiol Heart Circ Physiol 295:H1695–H1704

40. Kandasamy K, Prawez S, Choudhury S, More AS, Ahanger AA, Singh TU, Parida S, Mishra SK (2011) Atorvastatin prevents vascular hyporeactivity to norepinephrine in sepsis: role of nitric oxide and α1D-adrenoceptor mRNA expression. Shock 36:76–82

41. Reddy AK, Taffet GE, Madala S, Michael LH, Entman ML, Hartley CJ (2003) Noninvasive blood pressure measurement in mice using pulsed Doppler ultrasound. Ultrasound Med Biol 29:379–385

42. Kurtz TW, Griffin KA, Bidani AK, Davisson RL, Hall JE (2005) Recommendations for blood pressure measurement in humans and experimental animals. Part 2: Blood pressure measurement in experimental animals. Hypertension 45:299–310

43. Tsukamoto A, Serizawa K, Sato R, Yamazaki J, Inomata T (2015) Vital signs monitoring during injectable and inhalant anesthesia in mice. Exp Anim 64(1):57–64

44. Parasuraman S, Raveendran R (2012) Measurement of invasive blood pressure in rats. J Pharmacol Pharmacother 3:172–177

45. Kurowski SZ, Slavik KJ, Szilagyi JE (1991) A method for maintaining and protecting chronic arterial and venous catheters in conscious rats. J Pharmacol Methods 26:249–256

46. Ordodi VL, Mic FA, Mic AA, Toma O, Sandesc D, Paunescu V (2005) A simple device for invasive measurement of arterial blood pressure and ECG in the anesthesized rat. Timisoara Med J 55:35–37

47. Bardelmeijer HA, Buckle T, Ouwehand M, Beijnen JH, Schellens JH, van Tellingen O (2003) Cannulation of the jugular vein in mice: a method for serial withdrawal of blood samples. Lab Anim 37:181–187

48. McLachlan RS (1993) Suppression of interictal spikes and seizures by stimulation of the vagus nerve. Epilepsia 34:918–923

49. Hatton KW, McLarney JT, Pittman T, Fahy BG (2006) Vagal nerve stimulation: overview and implications for anesthesiologists. Anesth Analg 103:1241–1249

50. Choudhury S, Kannan K, Pule AM, Darzi SA, Singh V, Singh TU, Thangamalai R, Dash JR, Parida S, Debroy B, Paul A, Mishra SK (2015)

Combined treatment with atorvastatin and imipenem improves survival and vascular functions in mouse model of sepsis. Vascul Pharmacol 71:139–150. https://doi.org/10.1016/j.vph.2015.03.012

51. Kandasamy K, Choudhury S, Singh V, Addison MP, Darzi SA, Kasa JK, Thangamalai R, Dash JR, Kumar T, Sultan F, Singh TU, Parida S, Mishra SK (2016) Erythropoietin reverses sepsis-induced vasoplegia to norepinephrine through preservation of α1D-adrenoceptor mRNA expression and inhibition of GRK2-mediated desensitization in mouse aorta. J Cardiovasc Pharmacol Ther 21 (1):100–113. https://doi.org/10.1177/1074248415587968

52. Pfaffl MW (2001) A new mathematical model for relative quantification in real-time RT-PCR. Nucleic Acids Res 29:2002–2007

53. Livak KJ, Schmittgen TD (2001) Analysis of relative gene expression data using real-time quantitative PCR and the $2^{-\Delta\Delta ct}$ Method. Methods 25:402–408

Chapter 15

Methods to Study the Innate Immune Response to Sepsis

Wendy E. Walker

Abstract

This chapter describes techniques to measure the innate immune response in the mouse cecal ligation and puncture model of sepsis. The reader will learn how to perform retro-orbital bleeds to harvest serum from mice and learn how to perform peritoneal lavage to harvest cells and inflammatory mediators from this compartment. The enzyme-linked immunosorbent assay (ELISA) technique is described as a method to measure the levels of cytokines and chemokines in these fluids. Additionally, this chapter describes techniques to stain the cellular fraction of the peritoneal lavage with fluorescently labeled antibodies, and perform fluorescence activated cell sorting (FACS) to quantify macrophages and neutrophils in this compartment.

Key words Sepsis, CLP, Innate immune response, Inflammatory response, Retro-orbital bleed, Peritoneal lavage, Enzyme-linked immunosorbent assay (ELISA), Fluorescence-activated cell sorting (FACS), Cytokines, Macrophages, Neutrophils

1 Introduction

Sepsis occurs when an infection induces a systemic inflammatory response syndrome (SIRS). This is characterized by a "cytokine storm" that helps to fight the infection, but is paradoxically also detrimental to the host [1, 2]. In the laboratory, sepsis is frequently studied by performing cecal ligation and puncture (CLP) on rodents, a technique first described by Irshad Chaudry's group [3]. Following this procedure, commensal bacteria that normally reside in the gut gain access to the peritoneal compartment. This leads to a polymicrobial infection, starting with local peritonitis and progressing to bacteremia and SIRS. The related colon ascendens stent peritonitis (CASP) [4] and intraperitoneal cecal slurry injection [5] models induce sepsis in a similar fashion.

During sepsis, the host is exposed to pathogen associated molecular patterns (PAMPs) that activate an innate immune response. Additionally, damage of cells and tissues leads to the release of host-derived danger associated molecular patterns (DAMPs). Macrophages reside in the peritoneum and many other

Binu Tharakan (ed.), *Traumatic and Ischemic Injury: Methods and Protocols*, Methods in Molecular Biology, vol. 1717,
https://doi.org/10.1007/978-1-4939-7526-6_15, © Springer Science+Business Media, LLC 2018

tissues throughout the body. These cells express pattern recognition receptors (PRRs), including Toll-like receptors (TLRs), Retinoic acid-inducible gene I (RIG-I)-like receptors, and receptors of the cytosolic DNA-sensing pathway and inflammasome pathway. Activation of these pathways induces the release of proinflammatory cytokines and chemokines, both at the local site of infection and systemically (described by many groups including our own [6, 7]).

Chemokines orchestrate the recruitment of phagocytic cells to the primary site of infection, including neutrophils and additional macrophages. Early infiltration of neutrophils to the site of infection is beneficial during sepsis, as these phagocytic cells kill bacteria [8–10]. In contrast, prolonged activation of neutrophils and their recruitment to noninfected organs may damage these tissues and prove detrimental to the host [11, 12].

It is important to note that during sepsis, SIRS is accompanied by a compensatory anti-inflammatory response (CARS) [13] that compromises host defense against pathogens. For animals that survive the initial insult of severe sepsis, a persistent inflammatory and catabolic syndrome (PICS) may also develop [14].

Herein, we describe methods to measure the early innate immune response in mice following CLP-induced sepsis. To examine the systemic inflammatory response, animals are retro-orbitally bled and serum is prepared. To examine the local inflammatory response, animals are sacrificed at 18–24 h post-surgery and peritoneal lavage is performed. The enzyme-linked immunosorbent assay (ELISA) is performed on the serum and lavage samples to quantify inflammatory cytokines and chemokines. Additionally, peritoneal lavage cells are stained with fluorescently labeled antibodies against specific cell surface markers, followed by fluorescence activated cell sorting (FACS). This strategy allows the investigator to quantify neutrophils and macrophages in this compartment. In addition to their utility in the murine CLP model of sepsis, these methods are appropriate for use in the related CASP and cecal slurry models of sepsis, and may be adapted for use in other rodent species.

2   Materials

Equipment: This protocol requires standard lab equipment, including a benchtop centrifuge, microcentrifuge, and pipettmans.

2.1   Retro-Orbital Bleeds

1. Heparinized capillary tubes (i.e., Drummond Hemato-Clad mylar wrapped microhematocrit tubes, 75 mm).

2. Serum microtainers (i.e., BD microtainer SST, ref #365967).

3. Gauze pads.

4. Capillary tube pipette bulb or transfer pipette (to eject blood from capillary tubes).

5. 1.5 mL tubes (for storage of serum).

6. Proparacaine drops (optional, for topical ophthalmic anesthesia).

7. Eye lubricant ointment (i.e., Altalube ointment).

8. Anesthesia device (a precision isoflurane vaporizer or a drop anesthesia jar):

   (a) Isoflurane (for precision vaporizer).

   OR

   (b) 20–30% Isoflurane, 70–80% propylene glycol (for drop anesthesia jar).

2.2 Peritoneal Lavage

1. Dissection tools: forceps and scissors.

2. 5 mL Luer-lok syringes.

3. 23 G needles.

4. Phosphate-buffered saline (PBS), Ca, Mg-free.

5. 15 mL conical tubes.

6. Depending on the method of sacrifice, one of the following may be required:

   (a) An anesthesia device and isoflurane (*see* Subheading 2.1, **item 8**).

   OR

   (b) Carbon dioxide euthanasia chamber.

2.3 ELISA

1. The following ELISA reagents may be sold as a single set/kit or sold separately (i.e., BD OPTEIA ELISA sets, or eBioscience Ready-Set-Go! ELISA sets):

   (a) Capture antibody for the molecule of interest.

   (b) Detection antibody for the molecule of interest (biotinylated).

   (c) Streptavidin–horseradish peroxidase (HRP) (light sensitive).

   (d) Recombinant standard for the molecule of interest (aliquot and store at −80 °C upon receipt).

2. 96-well flat bottom ELISA plates (i.e., Nunc MaxiSorp plates).

3. Coating buffer: the appropriate solution may be indicated in the product sheet, or may be tested empirically in a test ELISA with a standard curve. Use within 2 weeks of preparation.

   (a) 0.2 M Sodium phosphate, pH 6.5: Mix 11.8 g Na_2HPO_4 + 16.1 g NaH_2PO_4 in 1 L H_2O, adjust pH to 6.5.

   OR

(b) Sodium carbonate, pH 9.5: 7.13 g NaHCO$_3$ + 1.59 g Na$_2$CO$_3$, adjust pH to 9.5.

4. Assay diluent: 10% heat-inactivated FBS, in PBS. Use within 2 weeks of preparation.

5. Capture antibody working solution: Capture antibody diluted ~1/250–1/1000 in assay diluent (*see* **Note 1**). Prepare immediately before use.

6. Working detector: Combine detection antibody (diluted ~1/250–1/1000) and streptavidin–HRP (diluted ~1/250) in assay diluent (*see* **Note 1**). Prepare immediately before use (light sensitive).

7. Wash buffer: 0.05% Tween 20 in PBS, Ca, Mg-free. Use within 2 weeks of preparation.

8. ELISA substrate reagent:

   (a) SuperAqua Blue (eBioscience, light sensitive).

   OR

   (b) TMB substrate (light sensitive).

   (c) Stop buffer: 1 M H$_3$PO$_4$ or 2 N H$_2$SO$_4$ (only required when using TMB substrate).

9. Adhesive plate sealers.

2.4 Enumerating and Staining Cells

1. Hemacytometer slide.

2. Ice cold staining buffer: 2% FBS in PBS, prepared within 1 week of use.

3. Antibodies against cell surface markers of interest, including CD11b, Ly6G and F4/80. These may be directly conjugated to a fluorochrome. Alternatively, a biotinylated antibody may be used in combination with a streptavidin–fluorochrome conjugate (available from eBioscience or BD bioscience).

4. Working solutions of antibody staining reagents (prepare immediately before use, *see* **Note 2**):

   (a) Single stain controls: dilute each antibody individually in staining buffer, at 10× the final dilution, i.e., make a 1/20 working solution for a 1/200 final dilution.

   (b) Antibody cocktail: dilute the antibodies in combination staining buffer, at 10× the final dilution, i.e., make a 1/20 working solution for a 1/200 final dilution.

   (c) Streptavidin–fluorochrome conjugate: dilute the streptavidin–fluorochrome conjugate in staining buffer, at 10× the final dilution, i.e., make a 1/50 working solution for a 1/500 final dilution.

5. Fixation buffer: 4% paraformaldehyde solution (i.e., IC fixation buffer, eBiosciences).

3 Methods

3.1 Retro-Orbital Bleeds

Retro-orbital bleed is a convenient method to obtain small samples of mouse blood. If performed correctly, this technique induces minimum discomfort to the animal and bleeding can be tightly controlled to avoid excess blood loss. Retro-orbital bleeds may be performed at serial time points to observe progression of the inflammatory response over time, without sacrificing the animal. Care should be taken to collect no more than 10% animal blood volume as the sum of all bleeds performed on a mouse within a 14-day period (*see* **Note 3**).

1. Anesthetize the mouse with isoflurane. The mouse should be under plane 2 (medium) anesthesia, unresponsive to toe pinch prior to commencing the procedure (*see* **Note 4**).

2. Optional: apply a drop of proparacaine solution to the eye for topical anesthesia.

3. Scruff the mouse firmly. This will tighten the skin around the face and pull the eyelids away from the eye socket.

4. Apply a heparinized capillary tube to the retro-orbital socket at the bottom front corner, between the orb of the eye and the lower eyelid. The tube should be held at a 45-degree angle to the animal's head (*see* Fig. 1). Apply gentle pressure while twiddling the tube back and forth (*see* **Note 5**). These small semicircular rotations will eventually ease the capillary tube into the space between the orb of the eye and the socket. Additional small rotations will puncture the conjunctival

Fig. 1 Retro-orbital bleeds. The mouse is scruffed to pull the skin away from the face, opening the eyelid and exposing the eye. The capillary tube is applied to the bottom front corner of the retro-orbital socket, between the orb of the eye and the lower eyelid, at a 45-degree angle to the animal's head

membrane, breaking through into the venous sinus and releasing blood.

5. Once blood begins to enter the capillary tube, ease back on the pressure applied to the capillary tube and loosen the pressure of the scruff hold on the animal. This will allow the blood to enter the capillary tube at a slow and steady rate. Once a sufficient volume has been collected, remove the capillary tube from the eye (*see* **Note 6**).

6. Release the scruff hold on the animal, allowing the skin to become more relaxed around the animal's face. Press the eyelid closed and then apply gentle pressure to the eyelid with a gauze pad, to staunch the bleeding. Bleeding should stop within 10–20 s. Apply an eye lubricant ointment to the edge of the eyelid and manually blink the eyelid a few times to spread the ointment across the eye. Ensure that all bleeding has stopped and that the animal is fully conscious and able to retain sternal recumbency before returning to the housing cage.

7. Flush the blood out of the capillary tube into the serum microtainer, using the capillary tube pipette bulb or transfer pipette (*see* **Note 7**). Close the microtainer tube.

8. Spin the serum microtainer tubes at $12,000 \times g$ for 2 min in a room-temperature centrifuge. The blood cells will migrate below the polymer gel in the microtainer tube, while the serum remains in a layer at the top. Carefully collect the serum layer and transfer into a 1.5 mL sample tube (*see* **Note 8**).

9. Store the samples at $-80\ ^{\circ}C$ (*see* **Note 9**).

3.2 Peritoneal Lavage

1. Sacrifice the animal (*see* **Note 10**).

2. Draw 3 mL PBS into a 5-mL Luer-Lok syringe and apply a 23 G needle to the end of the syringe. Expel any air bubbles from the syringe.

3. Perform a vertical incision in the skin adjacent to the midline surgical incision (*see* Fig. 2a). The incision should expose the underlying peritoneal membrane, but should not pierce it (*see* **Note 11**).

4. Pull up on the peritoneal membrane with the forceps and insert the needle just under the peritoneal membrane with the needle facing up (*see* Fig. 2b).

5. Inject 3 mL PBS into the peritoneal cavity. Allow the fluid to sit in the peritoneum for 2 min, while periodically massaging the abdomen.

6. Pull up on the peritoneal membrane with the forceps. Recover the PBS by making a tiny hole in the peritoneal membrane and sucking it out with a pipette (*see* Fig. 2c, d), or using a syringe (*see* **Note 12**). Transfer the lavage to a 15-mL falcon.

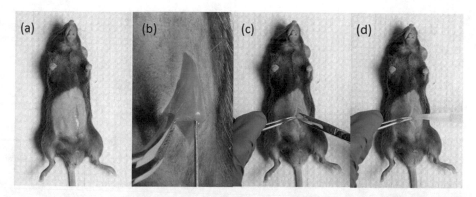

Fig. 2 Peritoneal lavage. (**a**) An off-center vertical incision is made in the mouse skin, leaving the peritoneal membrane intact (for a surgical mouse, this would be parallel to the surgical incision). (**b**) The peritoneal membrane is lifted away from the abdominal organs with the forceps. The needle is inserted just below the membrane, avoiding the organs, with the needle eye facing up. (**c**) After a 2-min incubation, with gentle massaging of the abdomen, the peritoneal membrane is once again lifted with the forceps and a small incision is made to accommodate the pipette tip. (**d**) The lavage fluid is withdrawn from the peritoneal cavity with the pipette

7. Spin the lavage solution at $600 \times g$ for 7 min. Collect the supernatant and store at $-80\ °C$.

8. Resuspend the pellet in 200 µl staining buffer, place the tube on ice and use for staining immediately.

3.3 ELISA

The ELISA was described by two independent groups in 1971 [15, 16]. This is a convenient method to measure the concentration of secreted proteins, such as cytokines and chemokines. Commonly measured proinflammatory cytokines include: interleukin (IL)-1, IL-6, IL-12, macrophage migration inhibitory factor (MIF), tumor necrosis factor (TNF)-α, as well as the type 1 interferons: IFN-α and IFN-β. Commonly measured anti-inflammatory cytokines include: IL-4, IL-10, and transforming growth factor beta (TGF-β). Commonly measured chemokines include: macrophage chemoattractant protein (MCP)-1, chemokine (C-X-C motif) ligand (CXCL)1 and CXCL2. Other soluble mediators may be measured in the CLP model, including myeloperoxidase (released by neutrophils that infiltrate the peritoneum) and serum amyloid P (the major mouse acute phase protein released by the liver into the bloodstream). Figure 3 shows representative serum cytokine data for IL-6, IL-12p40, and MCP-1 at 0, 5, and 19 h post-CLP and sham surgeries.

An ELISA troubleshooting chart is presented in Table 1.

1. The ELISA plate must be coated with capture antibody 1–3 days prior to the assay (*see* **Note 13**). Aliquot 100 µl of capture antibody working solution to an appropriate number of wells in a 96-well ELISA plate. Coat a sufficient number of

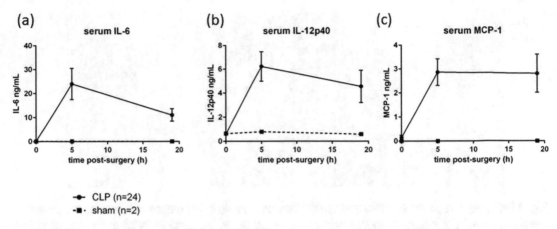

Fig. 3 Serum cytokines. Sample graphs showing the elevation in serum cytokines following CLP surgery vs. sham surgery. Graphs show (**a**) IL-6, (**b**) IL-12p40, (**c**) MCP-1

wells to run the samples and standards in duplicate or triplicate. Cover the plate with an adhesive plate sealer and store at 4 °C overnight (*see* **Note 14**).

2. On the day of the assay, wash the plate three times with 300 μl wash buffer per well (*see* **Note 14**).

3. Add 200 μl assay diluent to each well. Cover the plate with an adhesive plate sealer and incubate the plate at room temperature for 1 h (*see* **Note 15**). This step blocks nonspecific binding.

4. While the assay diluent is incubating, dilute the stock of recombinant cytokine or chemokine in assay diluent to make a standard curve. Typically a range of twofold dilutions from 2000 pg/mL down to 15.5 pg/mL is a suitable range (*see* Fig. 4a). Include a 0 pg/mL control (assay diluent only) (*see* **Note 16**).

5. Thaw the serum samples on ice and dilute in assay diluent as appropriate (*see* **Note 17**).

6. Wash the maxisorp plate three times with 300 μl wash buffer per well, and then blot dry.

7. Add 100 μl of standard or sample to the appropriate wells and incubate for 2 h at room temperature or at 4 °C overnight (*see* **Note 18**).

8. Wash the maxisorp plate three to five times with 300 μl wash buffer per well, and then blot dry (*see* **Note 19**).

9. Add 100 μl of working detector to each well (*see* **Note 20**). Cover with a plate sealer and incubate at room temperature for 1 h.

10. Wash the maxisorp plate five to seven times with 300 μl wash buffer per well, and then blot dry (*see* **Note 19**).

Table 1
ELISA troubleshooting

Problem	Possible causes
Technical replicates (i.e., duplicates, triplicates) yield very different absorbance values	– inaccurate pipetting – contamination of the plate – some wells dried out between solutions
No color development in the plate	– one or more steps was skipped – capture antibody was used in place of detection antibody – the detection antibody was for a different cytokine than the capture antibody (common mistake in a lab with multiple ELISA kits in identical boxes) – one or more ELISA reagents have degraded and need to be replaced – an incorrect type of 96-well plate was used for the assay (i.e., a tissue culture plate instead of a maxisorp plate)
Weak color development in samples and standards	– one or more ELISA reagents have degraded and need to be replaced – ELISA reagents were inappropriately diluted (use a higher concentration) – incorrect coating buffer was used (i.e., sodium phosphate vs. sodium carbonate)
Weak color development in samples, standards look OK	– sample is too dilute (concentration too low to measure) – sample is too concentrated (>1/20 concentration of serum will inhibit the ELISA assay) – sample has degraded (due to the unstable nature of cytokines and chemokines, serum samples should be frozen promptly at $-80\,^\circ$C after preparation and tested within 6 months)
Weak (or no) color development in standards, samples look OK	– incorrect dilution of the first (highest) standard – standards are old and need to be replaced (due to the unstable nature of cytokines and chemokines, standards only last ~6 months in $-80\,^\circ$C freezer) – incorrect standard was used, i.e., IL-6 instead of IL-12 (common mistake in a lab with multiple ELISA standards stored in the same box)
All wells turn an equally dark color rapidly	– detection antibody was used to coat the plate instead of capture antibody
Standard curve develops a dark color very rapidly	– standards are too concentrated (i.e., incorrect dilution of the highest standard)
High background (i.e., significant color development in the 0 pg/mL standard)	– insufficient washing of the plate – assay diluent was prepared incorrectly – wash buffer or PBS was used in place of assay diluent for the blocking step or sample dilution
Color develops too fast in the plate	– one or more reagents were too concentrated (i.e., capture or detection antibody)

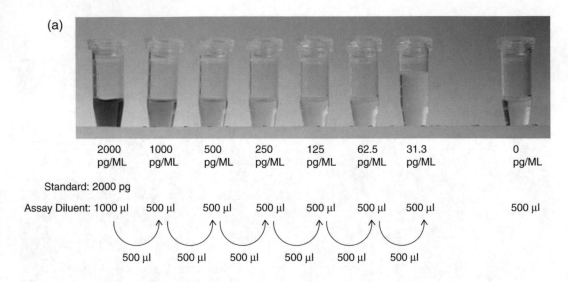

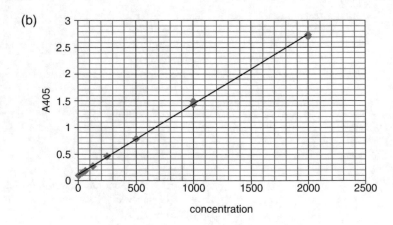

Fig. 4 ELISA standard curve. (**a**) Schematic of dilutions used to create the standard curve (please note that the blue color is for demonstration purposes only). (**b**) Plot of concentration vs. absorbance for the standard curve. Sample concentration is determined by comparing absorbance to the standard curve

11. Add ELISA substrate reagent. Follow step (**a**) or (**b**) according the reagent used (*see* **Note 21**).

(a) Add 100 μl Super AquaBlue ELISA substrate. Incubate the plate at room temperature for 30 min, or until blue color development can be visualized clearly in the plate. Read absorbance at 405 nm with a plate reader.

OR

(b) Add 100 μl TMB ELISA substrate per well. Incubate the plate at room temperature for 30 min, or until yellow color development can be visualized clearly in the plate. Add 100 μl stop buffer to each well. Read absorbance at 450 nm with a plate reader.

12. Create a standard curve by plotting the concentration of the standards against their absorbance (*see* Fig. 4b). Use a line of best fit or a point-to-point method to predict the concentration of the unknown samples (*see* **Note 22**). Multiply this concentration by the dilution factor used in **step 5** to determine the original concentration of the serum and lavage samples.

3.4 Enumerating and Staining Inflammatory Cells, with FACS Analysis

Staining peritoneal lavage cells with cell surface antibodies is a convenient method to quantify neutrophils (Ly6G+CD11b+CD64-F4/80-) and macrophages (Ly6G-CD11b+CD64+F4/80+) in this compartment. Additionally, cell surface staining may be used to visualize the maturation state and functionality of these cells. Figure 5 shows representative FACS plots for mice that were sacrificed 18 h after sham or CLP surgery and underwent peritoneal lavage. This figure also shows a comparison of lavage cells from interferon regulatory factor 3 (IRF3)-KO mice that have undergone CLP surgery, to highlight differences between genotypes that may be measured through FACS analysis.

1. Begin with the cell pellet obtained from the peritoneal lavage, resuspended in ice-cold staining buffer.

2. Count the cells with a hemacytometer slide (*see* **Note 23**).

3. Aliquot 0.5–2×10^6 cells into a 1.5 mL eppendorf tube for each staining reaction. Be sure to include tubes for an unstained control and a single stain control for each antibody used.

4. Wash the cells with 500 μl staining buffer and spin at $500 \times g$, 2 min in a microcentrifuge. Remove all but ~45 μl staining buffer. Repeat this step once, for a total of two washes (*see* **Note 24**).

5. Add 5 μl cell surface antibody cocktail working solution to the appropriate tubes, and single antibody control working solutions to the appropriate tubes. Add 5 μl staining buffer to the unstained control. Vortex each tube briefly to mix.

6. Incubate for 30–60 min on ice in the dark.

7. Wash with 500 μl staining buffer and spin $500 \times g$, 2 min. Remove all but ~45 μl liquid. Repeat once for a total of two washes.

(Note: **steps 8–10** are appropriate for antibodies that are conjugated to biotin and used in combination with a streptavidin–fluorophore conjugate. These steps should be skipped if the antibody is directly conjugated to the fluorophore).

8. Add 5 μl streptavidin–fluorophore conjugate working solution to the appropriate tubes and vortex briefly to mix.

9. Incubate for 15 min on ice in the dark.

10. Wash with 500 μl staining buffer and spin $500 \times g$, 2 min. Remove all but ~45 μl supernatant.

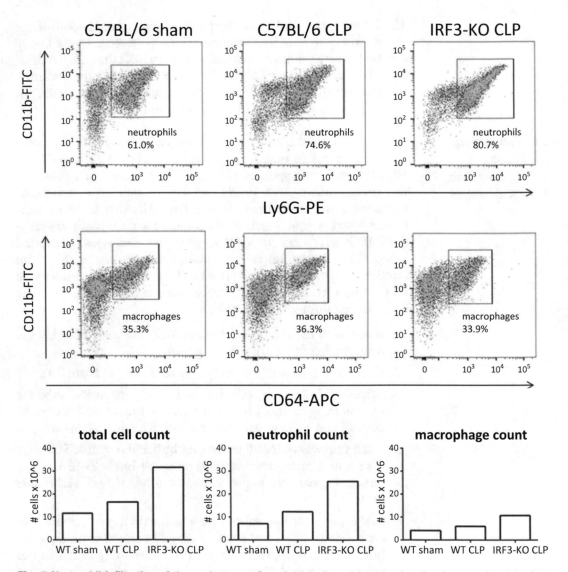

Fig. 5 Neutrophil infiltration of the peritoneum. Sample data for wild-type mice that have undergone sham surgery or CLP, as well as an IRF3-KO mouse that has undergone CLP surgery (the IRF3-KO strain exhibits increased neutrophil infiltration of the peritoneum (6)). (a) FACS plot of lavage cells stained with CD11b, Ly6G and CD64, identifying neutrophils and macrophages. (b) Cell counts in the peritoneum including: total cell count, neutrophil count (% neutrophils measured by FACS × total cells) and macrophage count (% marophages measured by FACS × total cells)

11. Add 200 μl IC fixation buffer.

12. Incubate the cells on ice 5–10 min (*see* **Note 25**).

13. Wash with 500 μl staining buffer and spin 500 × *g*, 2 min.

14. Remove all but ~45 μl supernatant and resuspend in 250 μl staining buffer.

15. Store the samples at 4 °C. Analyze the cells via FACS within 1 week (*see* **Note 26**).

4 Notes

1. The appropriate dilutions for the capture and detection antibodies and streptavidin–HRP may be indicated in the product sheet, or may be determined empirically by performing a test ELISA.

2. Prior to routine use, each antibody (and the streptavidin–fluorophore conjugate) should be titrated to determine an appropriate dilution for staining. The investigator is recommended to make serial dilutions of the antibody (i.e., 1/50, 1/100, 1/200, 1/400, 1/800) and test them for staining, using cells known to be positive for the relevant cell surface marker. This test will reveal a dose response indicating the minimum dose of antibody (or streptavidin–fluorophore) necessary to achieve full staining of the cells. The investigator is recommended to use the next dose up in the titration series for subsequent experiments (i.e., if 1/400 yields maximal staining, staining use 1/200 dilution for subsequent experiments). With this method, staining intensity will remain robust even when staining volumes vary slightly.

3. Serial small-volume retro-orbital bleeds may be performed on a single animal to monitor serum markers of inflammation over time. We typically bleed mice at 0 h, 4–6 h, and 18–24 h postsurgery. Once the investigator is proficient in the retro-orbital bleed technique, it may be preferable to alternate between the eyes to minimize discomfort to the animal. With practice both the left and right eye can be accessed with the dominant hand, or the investigator may become proficient in performing the technique with either hand. A maximum of 10% blood volume (roughly 1% of body weight) can be collected in a 14-day period. For a 20–25 g mouse, this corresponds to 200–250 μl. Larger volumes of blood may be collected from animals that will be sacrificed immediately thereafter.

4. Induction of anesthesia may be accomplished with a precision isoflurane vaporizer (set to ~2.0 isoflurane, 0.5 oxygen flow, or according to the manufacturer's specifications). Alternatively, a drop anesthesia chamber may be used, employing a mixture of 20–30% isoflurane, 70–80% propylene glycol. For the drop anesthesia chamber, anesthetic agent applied to a gauze pad, and a physical barrier (e.g., a screen) should be used to prevent skin contact with the animal. Care should be taken to ensure that the mouse is adequately anesthetized and completely unresponsive to toe pinch prior to commencing the retro-orbital bleed. The animal may be maintained on a small nose cone attached to the precision vaporizer during the retro-orbital

bleed. In the case of drop anesthesia, the procedure should be completed rapidly to ensure it is finished before the animal regains consciousness and the pain response returns. The investigator should monitor the animal carefully during induction of anesthesia, to prevent the animal from entering too deep of an anesthetic plane (this is often evidenced by an infrequent, gasping breath pattern). Septic animals are highly prone to death from anesthetic overdose.

5. Use gentle pressure when easing the tube into the socket, to avoid damage to the eye and to prevent a large and uncontrolled rush of blood, which may lead to excess animal blood loss.

6. Because sepsis induces a drop in blood pressure, the blood may flow less readily from the socket of a sick animal vs. a healthy one. Angling the end of the capillary tube downward may help blood enter the tube more readily. A proficient investigator may also lift the animal slightly with the hand holding the scruff, and angle its head downward during the bleeding procedure to facilitate entry of the blood into the capillary tube.

7. The bottom of the microtainer tube contains a polymer gel to facilitate separation of the serum. Care should be taken not to allow the capillary tube to touch this gel, which may clog the end and make it difficult to flush out the blood.

8. The serum microtainer should be held at an angle to allow the serum to pool in one corner for easy collection. Avoid touching the tip to the gel layer as the pipette tip may become clogged.

9. Note that cytokines and chemokines have a short half-life at room temperature. Therefore, serum should be processed and frozen quickly after blood collection. Storage at $-80\ ^{\circ}$C is recommended to preserve stability of these inflammatory mediators, although storage at $-20\ ^{\circ}$C is sufficient for short periods of time.

10. Animals may be sacrificed by CO_2 asphyxiation or by cervical dislocation under isoflurane anesthesia once the animal is unresponsive to toe pinch. In the latter case, the investigator should be careful not to apply excess pressure during the cervical dislocation, which may break the spine and cause bleeding into the peritoneal cavity, confounding the subsequent analyses. Death should be confirmed (i.e., by monitoring cessation of the heartbeat) prior to starting the lavage technique.

11. Exposure of the translucent peritoneal membrane will allow the investigator to visualize the proximal structures within the abdominal cavity. Visualization and pulling up on the peritoneal membrane will facilitate insertion of the needle into the cavity without piercing the intestines.

12. Collection of peritoneal lavage with a pipette through a small incision is technically easier than collection with a syringe, because the syringe needle can easily adhere to peritoneal fat and the intestines during collection. When collecting through an incision with a pipette, be careful to continue pulling up on the peritoneal membrane adjacent to the incision with the forceps throughout the procedure. Additionally, make the incision as small as possible to accommodate the pipette tip. Otherwise, the lavage fluid may flow out of the incision and be lost.

13. The plate may be incubated for up to 3 days with capture antibody at 4 °C, prior to performing the ELISA. Longer incubation periods are not recommended, as the assay sensitivity may be greatly reduced. Optionally, the plate may be incubated on an orbital shaker during all ELISA steps, to ensure mixing of solutions and increase exposure of the plastic to the solutions.

14. The plate may be washed with an automatic plate washer, or manual pipetting. A plate washer and a multichannel pipette are excellent time-saving devices, and may represent a sound investment for a lab that wishes to perform the ELISA technique on a regular basis. With the automatic plate washer, the waste solution may be dispensed into each row in turn, via a multichannel dispenser (i.e., Nunc Immuno Wash) and then sucked into a waste container connected that is connected to the same dispenser via a vacuum pump. With the automatic plate washer, it is recommended to allow the wash buffer to soak for 30 s–1 min for each wash prior to removal. With the manual pipetting technique, the liquid may be dispensed with a multichannel pipette and then disposed into a sink or large beaker via a rapid inversion of the plate with a flicking motion. The plate should subsequently be maintained in an inverted state until it is tapped dry on a stack of paper towels to remove all residual liquid. Regardless of which washing technique is used, the plate should be tapped dry on paper towels after the final wash, prior to addition of the next solution for the ELISA assay. Proceed immediately between the steps and do not allow the ELISA plate to sit dry for any period between washes or solutions.

15. The blocking step prevents nonspecific binding of irrelevant proteins to exposed areas of the plate.

16. ELISA is fairly tolerant of extended incubation times, as long as all of the wells are treated equally. For example, incubation with the sample, detection antibody and streptavidin–HRP may be extended by up to an hour at room temperature without compromising the experimental results.

17. The investigator may need to vary the dilution factor, depending on the marker being measured and the severity of sepsis that is induced. Please note that serum should be diluted at least 1/20, as a higher concentration of serum will inhibit binding of the cytokine to the capture antibody. Please note that samples should be replaced at −80 °C promptly to preserve their stability if they will be used for additional assays. The number of freeze–thaw cycles should be minimized.

18. Incubation at 4 °C overnight may improve the sensitivity of the assay and facilitate the detection of low concentrations or cytokines and chemokines.

19. The higher number of washes is recommended to reduce background when using an automatic plate washer. With manual washing, the lower number may suffice.

20. For most ELISAs, the detection antibody and streptavidin–HRP steps may be combined (i.e., these reagents may be codiluted in assay diluent and incubated together in a single step), to save time. However, for increased sensitivity, these agents may be used in series. In this case, repeat **steps 8** and **9**, first with detection antibody working solution, and then with streptavidin–HRP working solution.

21. In the author's experience, the Super AquaBlue reagent gives a better linear relationship between concentration and absorbance, while the TMB reagent has better sensitivity to detect very low sample concentrations.

22. Duplicate or triplicate samples should yield very similar absorbance values. The standard curve should be plotted on a graph and visually inspected to determine if the linear curve is suitable (i.e., if the points fall along a line). For example, the ELISA absorbance will begin to plateau at a certain concentration, in this case a point-to-point method may be more appropriate to determine the sample concentration (i.e., draw a separate line connecting the average values of adjacent standards). It is not appropriate to extrapolate the concentration for samples with an absorbance value above the standard range. In this case, the ELISA should be repeated using a higher sample dilution. ELISA standards should be included in every plate and should only be used to determine the concentration of samples run concurrently. This is because small variations in the procedure and conditions (including ambient temperature) may affect the absorbance values at a given concentration.

23. Dilute the sample ~1/20–1/100 to count with a hemacytometer slide.

24. With practice, the investigator can remove all but ~45 μl of the staining buffer from the cell pellet after the spin, judging the approximate volume by eye, and leaving the sample ready for

staining. It is not necessary to achieve an exact volume since the staining has been optimized to be robust across small variations in the antibody concentration (*see* **Note 2**).

25. Fixing for 5–10 min gently fixes the cells, and is optimal for maintaining sample integrity while preserving the cells for medium-term storage (up to 1 week). In the author's experience, longer fixation periods may lead to the generation an insoluble cell pellet in the subsequent centrifugation step, which is not suitable for FACS analysis. If the samples will be analyzed by FACS immediately after staining, the fixation step may be omitted.

26. Users unfamiliar with FACS analysis are referred to Flow Cytometry Protocols [17]. During FACS analysis, be sure to compensate for spectral overlap between adjacent fluorophores in adjacent channels (FITC vs. PE, PE vs. PERCP, PERP vs. APC, etc.).

References

1. Aikawa N (1996) Cytokine storm in the pathogenesis of multiple organ dysfunction syndrome associated with surgical insults. Nihon Geka Gakkai zasshi 97:771–777

2. Hack CE, Aarden LA, Thijs LG (1997) Role of cytokines in sepsis. Adv Immunol 66:101–195

3. Wichterman KA, Baue AE, Chaudry IH (1980) Sepsis and septic shock--a review of laboratory models and a proposal. J Surg Res 29:189–201

4. Zantl N et al (1998) Essential role of gamma interferon in survival of colon ascendens stent peritonitis, a novel murine model of abdominal sepsis. Infect Immun 66:2300–2309

5. Sam AD II, Sharma AC, Law WR, Ferguson JL (1997) Splanchnic vascular control during sepsis and endotoxemia. Front Biosci 2:72–92

6. Walker WE, Bozzi AT, Goldstein DR (2012) IRF3 contributes to sepsis pathogenesis in the mouse cecal ligation and puncture model. J Leukoc Biol 92:1261

7. Silver AC, Arjona A, Walker WE, Fikrig E (2012) The circadian clock controls toll-like receptor 9-mediated innate and adaptive immunity. Immunity 36:251–261

8. Craciun FL, Schuller ER, Remick DG (2010) Early enhanced local neutrophil recruitment in peritonitis-induced sepsis improves bacterial clearance and survival. J Immunol 185:6930–6938

9. Mercer-Jones MA et al (1997) Inhibition of neutrophil migration at the site of infection increases remote organ neutrophil sequestration and injury. Shock 8:193–199

10. Jin L, Batra S, Douda DN, Palaniyar N, Jeyaseelan S (2014) CXCL1 contributes to host defense in polymicrobial sepsis via modulating T cell and neutrophil functions. J Immunol 193:3549–3558

11. Ness TL, Hogaboam CM, Strieter RM, Kunkel SL (2003) Immunomodulatory role of CXCR2 during experimental septic peritonitis. J Immunol 171:3775–3784

12. Walley KR, Lukacs NW, Standiford TJ, Strieter RM, Kunkel SL (1997) Elevated levels of macrophage inflammatory protein 2 in severe murine peritonitis increase neutrophil recruitment and mortality. Infect Immun 65:3847–3851

13. Ward NS, Casserly B, Ayala A (2008) The compensatory anti-inflammatory response syndrome (CARS) in critically ill patients. Clin Chest Med 29:617–625. viii

14. Gentile LF et al (2012) Persistent inflammation and immunosuppression: a common syndrome and new horizon for surgical intensive care. J Trauma Acute Care Surg 72:1491–1501

15. Engvall E, Perlmann P (1971) Enzyme-linked immunosorbent assay (ELISA). Quantitative

assay of immunoglobulin G. Immuno-
chemistry 8:871–874

16. Van Weemen BK, Schuurs AH (1971) Immu-
noassay using antigen-enzyme conjugates.
FEBS Lett 15:232–236

17. Givan A (2004) In: Hawley T, Hawley R (eds)
Flow cytometry. Flow cytometry protocols.
Methods in molecular biology, vol 263.
Humana Press, Totowa, NJ, pp 1–31

Chapter 16

An Ovine Model for Studying the Pathophysiology of Septic Acute Kidney Injury

Yugeesh R. Lankadeva, Junko Kosaka, Roger G. Evans, and Clive N. May

Abstract

The development of acute kidney injury (AKI) is both a significant and independent prognostic factor of mortality in patients with sepsis, but its pathophysiology remains unclear. Herein, we describe an ovine model of sepsis evoked by the administration of live *Escherichia coli* in which there is hypotension, peripheral vasodilatation with a large increase in cardiac output; a similar hyperdynamic state to that commonly reported in humans. Interestingly, in this sheep model of sepsis, despite an increase in global kidney blood flow, there is a progressive reduction in renal function. Although renal hyperperfusion develops, renal tissue hypoxia due to redistribution of intrarenal blood flow may contribute to the pathogenesis of septic AKI. We have, therefore, developed a novel methodology to chronically implant combination probes to monitor intrarenal tissue perfusion and oxygen tension during the development of septic AKI in conscious sheep with hyperdynamic sepsis.

Key words Sepsis, Hypotension, Renal blood flow, Cardiac output, Acute kidney injury, Ischemia, Hypoxia

1 Introduction

Sepsis is the major cause of morbidity and mortality in intensive care units when associated with multiple organ dysfunctions [1, 2]. An organ most frequently showing sepsis related derangements is the kidney, with septic acute kidney injury (AKI) accounting for nearly 50% of cases of renal failure [2, 3]. The etiology of septic AKI is complex and poorly understood, but a reduced renal blood flow is thought to be central to the development of renal failure in sepsis [4].

We have established a model of hyperdynamic sepsis in conscious sheep that is induced via intravenous infusion of live *Escherichia coli* (*E. coli*). In this ovine model, sepsis is characterized by hypotension, tachycardia, peripheral vasodilatation, increased cardiac output, fever, hyperlactemia, and AKI [5, 6], a similar hemodynamic profile to that seen in patients with sepsis [7]. In this

Binu Tharakan (ed.), *Traumatic and Ischemic Injury: Methods and Protocols*, Methods in Molecular Biology, vol. 1717, https://doi.org/10.1007/978-1-4939-7526-6_16, © Springer Science+Business Media, LLC 2018

model of sepsis, there is a progressive development of AKI despite an increase in total renal blood flow (RBF) [6]. In agreement with this finding, RBF has been shown to be either preserved or to increase in patients with septic AKI [8, 9] and in other large animal models of hyperdynamic sepsis [10, 11], indicating that a reduction in total RBF is not a prerequisite for the development of septic AKI.

A possible mechanism contributing to septic AKI, in the face of renal hyperperfusion, is microcirculatory dysfunction leading to redistribution of intra-renal blood flow resulting in local tissue ischemia and hypoxia. However, the extent to which heterogeneity of intrarenal tissue perfusion influences oxygenation within the cortex and medulla has not been investigated in conscious large animal models of hyperdynamic sepsis. In an important advance we have developed a technique to chronically implant, into the renal cortex and medulla, fiber optic probes that we specifically designed and had custom made by the manufacturer (Oxford Optronix, Oxford, England). These probes remain in place and provide stable measurements for at least 3 weeks, without artifacts in freely moving sheep in their metabolic cages [12]. The probes provide real-time measures of tissue perfusion by laser Doppler flowmetry and importantly also tissue oxygen tension (tPO_2) by fluorescence lifetime oximetry [13–15].

In conscious sheep with hyperdynamic sepsis, we have recently demonstrated that there is a rapid onset of selective reductions in renal medullary tissue perfusion and tissue oxygen tension within the first 4 h of sepsis, while cortical perfusion and oxygenation were preserved [16]. These findings suggest that medullary tissue hypoxia due to intra-renal blood flow redistribution may be one of the contributing factors for AKI during sepsis.

2   Materials

All experiments were conducted in 1–2-year-old pure bred female Merino ewes (35–40 kg), housed in individual metabolic cages. Prior to surgery, animals were allowed 8–10 days of acclimatization to laboratory conditions and human contact. Sheep were fed a diet of oaten chaff (800 g/day) and were allowed free access to water. All experimental procedures were approved by the Animal Ethics Committee of the Florey Institute of Neuroscience under guidelines laid down by the National Health and Medical Research Council of Australia.

2.1 Anesthetic, Antibiotic, and Analgesic Agents

1. Sodium thiopentone (15 mg/kg, Jurox PTY LTD, NSW, Australia) (*see* **Note 1**).

2. Isoflurane (Abbott Australasia PTY LTD, NSW, Australia) (*see* **Note 2**).

3. Procaine penicillin (900 mg, Ilium Propen, Troy Laboratories, Smithfield, NSW or Mavlab, QLD, Australia).

4. Flunixin meglumine (1 mg/kg, Troy Laboratories, NSW or Mavlab, QLD, Australia).

5. Gentamycin (150 mg/mL, Troy Laboratories, Glendenning, NSW, Australia).

2.2 Equipment for Monitoring Systemic and Renal Hemodynamics

1. Transit-time flow probe (20 mm, Transonic Systems, Ithaca, NY, USA).

2. Transit-time flow probe (4 mm, Transonic Systems, Ithaca, NY, USA).

3. Custom-built fiber-optic probes (450 μm outer diameter, CP-004-001 Oxford Optronix, Oxford, England).

4. Flow meter (T206, Transonic Systems, Ithaca, NY, USA).

5. OxyLite 2000 and OxyFlo monitors (Oxford Optronix, Oxford, England).

6. CED micro 1401 and Spike 2 software (Cambridge Electronic Design, Cambridge, England).

7. Tygon catheters (ID 1.0 mm, OD 1.5 mm, Cole-Parmer, Boronia, Australia).

8. Polythene catheters (ID 1.19 mm, OD 1.7 mm; Portex®, Smiths Medical International Ltd. Hythe, Kens UK).

9. Polythene catheters (ID 0.58 mm, OD 0.96 mm; Portex®; Smiths Medical International Ltd. Hythe, Kens UK).

10. Pressure transducer (ITL Healthcare, Chelsea Heights, VIC, Australia).

11. Heparin (Pfizer, West Ryde, NSW, Australia).

12. Foley bladder catheter (size 12, 30 cm^3, Euromedical, Malaysia).

13. Fractional urine collector (LKB Instruments, Victoria, Australia).

14. Gemini PC-1 volumetric infusion controller (IMED Corporation, CA, USA).

15. Blood gas analyzer (ABL System 625, Radiometer Medical, Copenhagen, Denmark).

2.3 Escherichia coli and Broth Reagents

1. *Escherichia coli* stock solution (isolated from blood cultures from a patient who recovered from septic shock (Austin Health, Melbourne, Australia) stored in 100 μl aliquots in sterile microtubes at −80 °C).

2. To prepare broth, weigh 10 g tryptone (Oxoid LTD, Hampshire, UK), 5 g yeast extract (Oxoid LTD, Hampshire, UK)

and 10 g NaCl (Merck, Damstadt, Germany), and dissolve in 1000 mL distilled water (*see* **Note 3**).

3. VITEK colorimeter (Hach Company, Colorado, USA).

4. Incubator (Orbital shaker, Thermo Scientific, Georgia, USA).

3 Methods

Prior to experimentation, all sheep undergo two aseptic surgical procedures under general anesthesia. Food and water are removed for 15–20 h prior to surgery to prevent ruminal tympany (bloat) and to reduce the risk of regurgitation and aspiration. All surgical procedures are conducted under aseptic conditions; all instruments are autoclaved and surgeons wear sterile gowns and gloves. The incision site is shaved with a small animal clipper and thoroughly scrubbed with Riodine (Povidone-Iodine, Perrigo, WA, Australia), following which the animal is covered with sterile drapes leaving only a minimum area exposed for incisions. For all surgical procedures, sheep are treated with intramuscular antibiotics (900 mg, procaine penicillin) at the start of surgery and at 24 and 48 h postoperatively. Analgesia is maintained with intramuscular flunixin meglumine (1 mg/kg) given at the start of surgery and at 4 and 24 h postsurgery.

3.1 Surgical Preparation of a Carotid Arterial Loop

1. The aim of this procedure is to exteriorize the left carotid artery into a skin fold to allow easy access for arterial cannulation and blood sampling.

2. Following anesthesia (*see* **Notes 1** and **2**), position the animals in dorsal recumbence and shave the neck area.

3. Place the neck in a ventrolateral position and make a skin incision ~12 cm in length parallel to the jugular vein and blunt dissect the skin free from underlying muscles and tissues.

4. Locate the carotid artery, and separate it from all the surrounding connective tissue, taking care not to damage the vagus and sympathetic nerves.

5. After a suitable length of the artery (~6 cm) has been freed from the surrounding tissues or vessels, make a second skin incision (~10 cm) in length in parallel to the first incision, producing a strip of skin about 2.5 mm wide that is wrapped around the carotid artery enclosing the vessel into a tube.

6. After wrapping the skin strip around the artery, neatly oppose the edges, and then close using a continuous suture lock technique.

During the same surgery, after the preparation of a carotid arterial loop, position the animal in right lateral recumbence and perform a left-sided thoracotomy for the implantation of a transit-time flow probe around the pulmonary artery to measure cardiac output (CO).

3.2 Cardiac Surgery

1. Perform a left flank skin incision above the fourth left rib. Strip the periosteum, and then remove the fourth rib.

2. Open the pericardium and carefully separate the pulmonary artery from the surrounding tissue.

3. Gently, place the transit-time flow probe (20 mm) around the vessel and take care to ensure a good alignment with the artery.

4. Insert a drain tube and then close the intercostal muscle and skin in layers and tunnel the flow probe cable subcutaneously and exteriorize near the thoracic spine.

5. Switch the anesthetic off and remove the endotracheal tube once corneal, laryngeal and muscle reflexes had been established.

6. Once the animal is breathing normally on its own, gently move it to an individual metabolic cage and allow recovery for 3 weeks. Monitor food and water intake daily and regularly check the wounds to ensure healing.

7. Maintain chest drainage for 4 h post-surgery, after which the drain is removed.

Following a recovery period of 2–3 weeks, an additional surgical procedure is performed to equip sheep with a renal artery flow probe, a renal vein catheter and combination fiber-optic probes into the renal cortex and medulla.

3.3 Renal Surgery

1. Following anesthesia, position the animals in right lateral recumbence.

2. Perform a paravertebral laparotomy to locate the left renal artery and kidney in the retroperitoneal space (*see* **Note 4**).

3. Carefully isolate the renal artery from the surrounding connecting tissue and fat taking care not to damage the renal nerves. Place a transit-time flow probe (4 mm) around the vessel, and then wrap with a silicone sheet. Suture both the probe and the film with polyamide suture after having ensured that the probe is in good alignment with the artery (*see* **Note 5**). This technique has been previously validated to provide an accurate index of RBF in conscious sheep [17].

4. Isolate the left renal vein, taking care not to damage the renal nerves. Insert a Tygon catheter (ID 1.0 mm, OD 1.5 mm) into the renal vein and secure in place with a purse string suture for blood sampling.

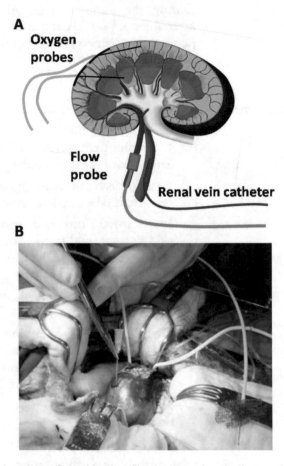

Fig. 1 Implantation of combination fiber-optic probes in the renal cortex and medulla to measure perfusion and oxygenation in conscious sheep. (**a**) location of the cortical and medullary probes and the sites of the renal transit-time flow probe and renal vein catheter. (**b**) photograph showing the implantation of fiber optic probes in an anesthetized sheep. Figure reproduced from Calzavacca et al. (2015) *Am J Physiol Regul Integr Comp Physiol*, 308: R832-R839, with permission [12]

5. Expose the left kidney from the surrounding tissue and fat. Insert the custom-built fiber-optic probes (450 μm outer diameter) with 20 mm of optical fiber extending from the outer sheath into the medulla and cortex along tracks prepared by previous insertion of a 25-gauge needle (514 μm outer diameter; Fig. 1) (*see* **Notes 6** and 7).

6. Insert each probe 20 cm into the kidney, one at an acute angle (~10°), so that the probe tip is positioned within the cortex (~3–4 mm below the renal capsule) and one at a ~60° angle, so that the tip is positioned within the medulla (6–10 mm below the renal cortex (Fig. 1) (*see* **Note 8**).

7. Confirm the positions of the probe tips at *post mortem* at the end of experimentation.

8. Tunnel the renal flow probe cable and the two renal vein catheters subcutaneously and exteriorize near the lumbar spine. Exteriorize the two fiber optic probe cables through the surgical wound. Close the longitudinal and transverse muscles in layers with a running suture, and close the skin with a continuous suture lock technique (*see* **Note 9**).

9. After the surgical implantation of all probes, insert a Tygon catheter (ID 1.0 mm, OD 1.5 mm) ~20 cm into the carotid artery for measurement of arterial pressure and collection of blood. Secure the catheter in place using a purse string suture. Insert two polythene catheters into the left jugular vein: one for the infusion of saline (ID 1.19 mm, OD 1.7 mm) and one for administering *E. coli* (0.58 mm, OD 0.96 mm) (*see* **Note 10**).

10. Following cannulation, switch the anesthetic off, and once the animal has regained consciousness, gently move it into an individual metabolic cage for recovery, as previously described.

Animals are allowed at least 5 days recovery from the renal surgery procedure in order to minimize any effects of surgical stress, prior to experimentation.

3.4 Animal Preparation

Carry out all experiments in conscious sheep to avoid the confounding effects of anesthesia, which has been shown to inhibit the increases in CO and RBF in sepsis.

1. The day prior to the start of experiments, insert a Foley bladder catheter (size 12) and connect this to the fraction collector for collection of urine at hourly intervals.

2. Connect the arterial line to a pressure transducer to allow measurement of mean arterial pressure (MAP) and heart rate (HR). Connect the pulmonary and renal probes to a flowmeter for monitoring CO and RBF (Figs. 2 and 3).

3. Connect the fiber optic probes to OxyLite™ 2000 and Oxy-FLo™ monitors for continuous monitoring of renal cortical and medullary laser-Doppler flux (arbitrary units), tissue oxygen tension (PO_2; mmHg) and tissue temperature (°C) (Fig. 3).

4. Continuously record analog signals for MAP, HR, CO, RBF, cortical and medullary laser-Doppler flux, tissue PO_2, temperature in conscious sheep at 100 Hz on a computer using a CED micro 1401 interface and Spike 2 software (Cambridge Electronic Design, Cambridge, UK) (Fig. 3) (*see* **Note 11**).

5. Calculate stroke volume (SV), total peripheral conductance (TPC) and renal vascular conductance (*see* **Note 12**).

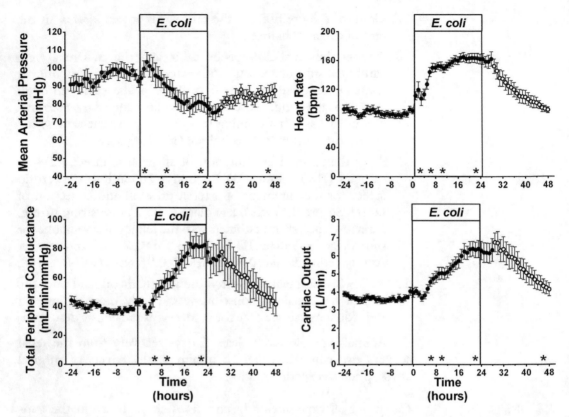

Fig. 2 Changes in mean arterial pressure, heart rate, total peripheral conductance and cardiac output in conscious sheep during a 24-h baseline period, a 24 h infusion of *Escherichia coli*, and a 24-h recovery period. Data are mean ± SEM. *$p < 0.01$ compared with baseline; $n = 8$ up to 28 h (filled circles) and then $n = 6$ (open circles). Figure reproduced from Calzavacca et al. (2015) *Crit Care Med*, 43: e431–e439, with permission [16]

3.5 Experimental Protocol

1. The day following the insertion of the bladder catheter, start measuring cardiovascular and renal variables over a 24 h baseline period (Figs. 2 and 3).

2. Following 24 h of baseline measurements, induce sepsis by intravenous infusion of live *E. coli* (6.8×10^7 colony forming units [CFU]/kg over 20 min in normal saline) as a bolus, followed by a continuous infusion for 24 h (3×10^7 CFU/kg/h) (Figs. 2 and 3) (*see* **Note 13**).

3. Administer fluid replacement intravenously with normal saline (154 mM NaCl, at 1 mL/kg/h) using a Gemini PC-1 volumetric infusion controller from the start of the baseline period until the end of the intervention period (72 h) to prevent hypovolemia.

4. Collect arterial and renal venous blood samples at baseline, just prior to *E. coli* infusion, and then at 4, 8, 12, and 24 h of sepsis, and at the end of the recovery period for measurement of blood oximetry and lactate (ABL System 625). Calculate renal

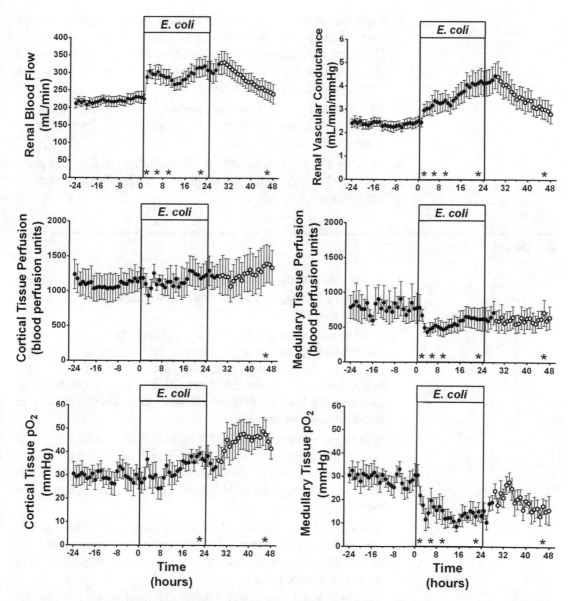

Fig. 3 Changes in renal blood flow and conductance, cortical and medullary tissue perfusion and oxygenation in conscious sheep during a 24-h baseline period, a 24 h infusion of *Escherichia coli*, and a 24-h recovery period. Data are mean ± SEM. *$p < 0.01$ compared with baseline; $n = 8$ up to 28 h (filled circles) and then $n = 6$ (open circles). Figure reproduced from Calzavacca et al. (2015) *Crit Care Med*; 43: e431–e439, with permission [16]

oxygen delivery (RDO_2), renal oxygen consumption (RVO_2) and renal oxygen extraction ratio (*see* **Note 14**).

5. Simultaneously, collect arterial blood and urine samples for measurement of creatinine and sodium. Calculate creatinine clearance, as a measure of glomerular filtration rate, and fractional sodium excretion (*see* **Note 15**).

6. After induction of sepsis for 24 h, stop the *E. coli* infusion, and give animals an intramuscular injection of gentamycin (150 mg) and monitor the sheep over a 24 h recovery period (Figs. 2 and 3). If required, it is possible to continue the infusion of *E. coli* up to 48 h [6].

7. At the end of the 24 h recovery period, humanely euthanize the animals with an overdose of pentobarbital intravenously (Lethabarb, 100 mg/kg).

8. Remove the kidneys at autopsy in order to confirm the positions of the fiber-optic probe tips and to visualize any signs of hematoma.

4 Notes

1. Induce anesthesia with intravenous injection of sodium thiopentone (15 mg/kg) to enable endotracheal tube placement (cuffed size 10).

2. Following tracheal intubation, connect the animals to a mechanical ventilator and maintain anesthesia with oxygen/air/isoflurane (end-tidal isoflurane, 1.5–2.0% v/v). Alter fractional inspired oxygen to maintain PaO_2 at approximately 100 mmHg and control ventilation to maintain $PaCO_2$ at approximately 40 mmHg.

3. Mix tryptone, yeast, and NaCl in 1000 mL distilled water in bottle, and autoclave (125 °C, fluid cycle for 1.5 h) with cap half unscrewed and allow to cool overnight. Divide the broth into 50 mL aliquots in sterilized falcon tubes and store at −4 °C.

4. Make a (~10 cm) paravertebral skin incision, then blunt dissect the underlying transverse and longitudinal muscle layers to reveal the peritoneum and gently ease the peritoneum away from the muscle layer.

5. Collect (~5 mL) of blood from the animal at the start of the surgery and allow clotting in a sterile container. Add these blood clots to fill the gap between the flow probe and the artery. The blood clots and silicone sheet prevent movement of fat into the probe window and the silicon sheet also contributes to immobilizing the probe in the proper position.

6. Each combination probe consists of a dual-fiber-laser-Doppler probe for estimating tissue perfusion by measurement of laser-Doppler flux, a single-fiber fluorescence optode for measuring tissue oxygen tension, and a thermocouple for measuring tissue temperature.

7. To prepare flexible sheet to secure the probes in the kidney, apply PVC glue thinly over a single layer of gauze, and allow

to dry overnight. Before insertion of probes, join the outer sheath of each probe with PVC glue to the flexible sheet (2.0 × 1.0 cm).

8. Cut out four triangular segments from the flexible sheet, so that it fits flush to the kidney surface before being glued to the kidney capsule with cyanoacrylate adhesive.

9. Fasten the fiber optic and flow probe connecting leads to the wool on the sheep's back, to prevent tangling and chewing of lines.

10. Securely fasten the arterial and venous lines to the wool on the sheep's back and connect to a pressurized bag of heparinized saline (10 IU heparin/mL) and continuously infused at 3 mL/h to maintain patency of the catheters.

11. Run all flow probe and fiber-optic cable connecting leads at the back of the cage over a freely revolving tube (10 cm diameter), allowing continuous monitoring of cardiovascular and renal variables in freely moving sheep in their metabolic cages.

12. Calculate stroke volume (SV) as CO/HR, total peripheral conductance (TPC) as CO/MAP adrenal vascular conductance (RVC) as RBF/MAP.

13. Measure the concentration of *E. coli* for infusion using a VITEK colorimeter. To reconstitute *E. coli*, add 100 µL of stock solution into 10 mL of broth and incubate overnight for 12–15 h. Following incubation, dilute *E. coli* with 30–35 mL of broth until the desired concentration is achieved. A 30% calorimeter reading is established to have 1.4×10^9 CFU/mL via microbiology examination.

14. Calculate renal O_2 delivery (RDO_2), O_2 consumption (RVO_2) and O_2 extraction ratio as follows: RDO_2 = (arterial O_2 content/100) × RBF; RVO_2 = [(arterial − renal venous O_2 content)/100] × RBF; O_2 extraction ratio = [(arterial − renal venous O_2 content)/(arterial O_2 content)] × 100.

15. Calculate creatinine clearance as Creatinine$_{Urine}$/Creatinine$_{Plasma}$ × Urine$_{Volume}$/Time, and fractional sodium excretion as Sodium$_{Urine}$/Sodium$_{Plasma}$ × Creatinine$_{Plasma}$/Creatinine$_{Urine}$ × 100.

Acknowledgment

This work was supported by grants from the National Health and Medical Research Council of Australia (NHMRC, 454615, 1009280, 1050672), and by funding from the Victorian Government Operational Infrastructure Support Grant. YRL was supported by a Postdoctoral Fellowship by the National Heart Foundation of Australia (NHF, 100869).

References

1. Mori T, Shimizu T, Tani T (2010) Septic acute renal faliure. Contrib Nephrol 166:40–46
2. Bagshaw SM, Uchino S, Bellomo R et al (2007) Septic acute kidney injury in critically ill patients: clinical characteristics and outcomes. Clin J Am Soc Nephrol 2:431–439
3. Bagshaw SM, George C, Bellomo R et al (2008) Early acute kidney injury and sepsis: a multicentre evaluation. Crit Care 12:R47–R47
4. Schrier RW, Wang W (2004) Acute renal failure and sepsis. N Engl J Med 351:159–169
5. Di Giantomasso D, May CN, Bellomo R (2003) Vital organ blood flow during hyperdynamic sepsis. Chest 124:1053–1059
6. Langenberg C, Wan L, Egi M et al (2006) Renal blood flow in experimental septic acute renal failure. Kidney Int 69:1996–2002
7. Dellinger RP, Levy M, Rhodes A et al (2013) Surviving sepsis campaign: international guidelines for management of severe sepsis and septic shock, 2012. Intensive Care Med 39:165–228
8. Brenner M, Schaer GL, Mallory DL et al (1990) Detection of renal blood flow abnormalities in septic and critically ill patients using a newly designed indwelling thermodilution renal vein catheter. Chest 98:170–179
9. Rector F, Goyal S, Rosenberg I et al (1973) Sepsis: a mechanism for vasodilation in the kidney. Ann Surg 178:222–226
10. Weber A, Schwieger IM, Poinsot O et al (1992) Sequential changes in renal oxygen consumption and sodium transport during hyperdynamic sepsis in sheep. Am J Physiol 262:F965–F971
11. Chvojka J, Sykora R, Krouzecky A et al (2008) Renal haemodynamic, microcirculatory, metabolic and histopathological responses to peritonitis-induced septic shock in pigs. Crit Care 12:R164–R164
12. Calzavacca P, Evans RG, Bailey M et al (2015) Long-term measurement of renal cortical and medullary tissue oxygenation and perfusion in unanesthetized sheep. Am J Physiol Regul Integr Comp Physiol 308:R832–R839
13. Griffiths J, Robinson S (1999) The OxyLite: a fibre-optic oxygen sensor. Br J Radiol 72:627–630
14. Leong CL, O'Connor PM, Eppel GA et al (2008) Measurement of renal tissue oxygen tension: systematic differences between fluorescence optode and microelectrode recordings in anaesthetized rabbits. Nephron Physiol 108:p11–p17
15. O'Connor P, Anderson W, Kett M et al (2007) Simultaneous measurement of pO2 and perfusion in the rabbit kidney in vivo. Adv Exp Med Biol 599:93–99
16. Calzavacca P, Evans RG, Bailey M et al (2015) Cortical and medullary tissue perfusion and oxygenation in experimental septic acute kidney injury. Crit Care Med 43:e431–e439
17. Bednarik JA, May CN (1995) Evaluation of a transit-time system for the chronic measurement of blood flow in conscious sheep. J Appl Physiol 78:524–530

Chapter 17

An In Vitro Model of Traumatic Brain Injury

Ellaine Salvador, Malgorzata Burek, and Carola Y. Förster

Abstract

Traumatic brain injury (TBI) is a significant problem causing high mortality globally. Methods to increase possibilities for treatment and prevention of secondary injuries resulting from the initial physical insult are thus much needed. TBI affects the central nervous system (CNS) and the neurovascular unit as a whole in numerous ways but one of the primarily compromised components is the blood–brain barrier (BBB).

In this chapter, we present a detailed procedure on how stretch injury and oxygen–glucose deprivation (OGD) are applied to brain microvascular endothelial cells of the BBB in order to replicate the actual impact they receive during TBI.

Key words In vitro model, Traumatic brain injury, Stretch injury, Murine cerebrovascular endothelial cells, cEND

1 Introduction

The study of traumatic brain injury (TBI) requires a model system that allows a close representation of the actual event. Various in vivo and in vitro models in different animals are available [1]. Nonetheless, in vitro models enable better manipulation and provide easier handling for experimental purposes. Hence, they are the preferred model for investigation of mechanisms involved in the sequelae of events following injury. It is believed that in vitro models of mechanical injury are a valuable tool for the study of the cellular consequences of TBI. In addition, they are useful for evaluation of potential therapeutic strategies of TBI [2]. Among the currently developed methods, the use of stretch in cultured cells via a stretch-inducing apparatus has gained acceptance in many laboratories. The use of stretch was first developed by Ellis and coworkers [3] through the establishment of the cell-injury controller (CIC) machine. Stretch has been used to injure various cell types in vitro [3–7]. Administration of stretch to cultured murine brain endothelial cells has also been employed in studying the cellular and

Binu Tharakan (ed.), *Traumatic and Ischemic Injury: Methods and Protocols*, Methods in Molecular Biology, vol. 1717, https://doi.org/10.1007/978-1-4939-7526-6_17, © Springer Science+Business Media, LLC 2018

molecular mechanisms involved in TBI at the blood–brain barrier (BBB) [8].

The cell injury controller is a device that induces injury to cultured cells through delivery of a controlled pulse of compressed gas to the cells in culture medium [7]. The mechanism behind its function is modelled after the hypothesis that TBI occurs as a function of the strain and strain rate experienced by the central nervous system (CNS) during a traumatic insult. In using the CIC device, cells cultured on plates with elastic membrane bottom are required. The machine applies pressure to the wells through a controlled flow of gas. The pressure generates a biaxial stretch injury whose severity can be manually set and controlled.

Meanwhile, oxygen–glucose deprivation (OGD) is the method of choice in replicating ischemia in vitro. Reduction of blood supply during ischemia results in a decrease of oxygen as well as glucose in the brain. Ischemia generally proceeds after the occurrence of head injury and is thus a part of the sequelae of events during TBI. Thus, its combination with stretch would better replicate the actual events occurring during TBI.

An important aspect of a suitable in vitro model system for TBI is the cell line of choice. Although various cell types ranging from astrocytes, neurons and endothelial cells [3, 5, 6] generated from murine, bovine and human sources have been employed, favored for the investigation of alterations in the BBB as a result of TBI are brain endothelial cells. In our laboratory, the brain microvascular endothelial cells cEND have been generated as an in vitro BBB model [9, 10] and are employed for our TBI studies in vitro. TBI leads to BBB disruption and the resulting increase in cerebrovascular permeability contributes to vasogenic brain edema and inflammation [11, 12]. These, in turn, can significantly influence TBI outcome. Therefore, studying the changes that occur at the BBB during TBI through an in vitro model system using cerebrovascular endothelial cells could be useful in finding strategies to deal with both the primary and secondary events that take place during actual TBI.

The protocol provided herein is subdivided into four sections, namely: preparation and culture of cEND cells, differentiation of cEND cells prior to stretch, stretch-induced injury of the cells, and oxygen–glucose deprivation (OGD) of the cells.

2   Materials

The following materials are needed for conducting in vitro TBI experiments using stretch:

2.1   Equipment

1. Stretch-injury machine (e.g., Cell Injury Controller (CIC) II (FlexCell International Corp, USA).

2. Cell culture incubator.

3. Hypoxic work station or incubator (e.g., Ruskinn Invivo2 Hypoxia Work Station, The Baker Company, USA).

4. Heat Plate (Optional).

5. Pipettor.

2.2 Biological Material

1. Cerebrovascular endothelial cells (e.g., cEND).

2. Collagen-coated 6-well flexible bottomed culture plates (e.g., BioFlex, FlexCell International Corp, USA).

2.3 Buffers, Solutions, and Media

1. Endothelial cell culture medium (supplemented with fetal bovine serum, endothelial cell growth supplement, endothelial growth factor, hydrocortisone, heparin, L-glutamine, and antibiotic–antimycotic solution).

2. Serum-stripped fetal calf serum.

3. Phosphate buffered saline (PBS).

4. Trypsin–EDTA solution.

5. Viability Stain (e.g., Image-iT® DEAD Green™ Viability Stain, Invitrogen).

6. LDH Release Assay Kit (e.g., Cytotoxicity Detection Kit Plus, Roche).

3 Methods

An important factor for the success of the in vitro TBI experimental procedure described here is the condition of the cultured cells. The cells should remain viable and morphologically suitable. Much care should be given in the proper maintenance of the cells for use in the experiments. In so doing, one is assured that the cells remain intact after stretch. The following outlines a detailed methodology of carrying out stretch-induced injury in cultured cEND cells followed by oxygen–glucose deprivation (OGD). A method of assessing the viability of the cells after application of stretch and OGD is also described. Protein and mRNA expression analyses methods subsequent to the treatment are not described herein. However, standard methodologies are available in published literature for reference (*see* Fig. 1).

3.1 Preparation and Culture of cEND Cells

1. Prior to experiment, cultivate cerebromicrovascular endothelial cells (cEND) or any other brain microvascular endothelial cells of choice in T75 culture flask until confluent. Medium should be changed twice a week (*see* **Note 1**).

2. Upon reaching confluence, cells need to be seeded into 6-well flexible bottomed culture plates precoated with collagen. To do

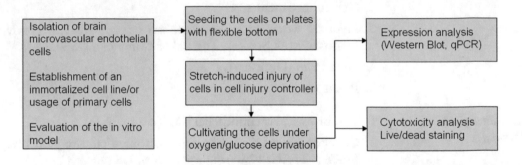

Fig. 1 Schematic representation of the described protocol. We describe in detail the seeding of the cells on flexible-bottomed plates, applying stretch to the cells and culturing the cells under oxygen/glucose deprivation conditions. We refer to other publications for Western blotting and qPCR protocols. We describe the assessment of cell injury using live/dead staining and cytotoxicity assay

this, first, wash the cells with phosphate buffered saline (PBS). Next, remove the PBS and trypsinize the cells with 2 mL warm trypsin-EDTA solution (*see* **Note 2**).

3. Allow the cell layer to be dispersed by incubating at 37 °C for 5 min (*see* **Note 3**).

4. Next, add 5 mL culture medium containing fetal calf serum to deactivate trypsin into the cells. Tap the flask gently with the hand several times to facilitate cell detachment.

5. View cells under the microscope to ensure complete detachment from flask.

6. Afterward, pipette medium with detached cells up and down for homogenous distribution of cells in the medium. Swirl the flask to mix the cell suspension.

7. Take 100 μl of the cell suspension for cell counting. Put appropriate amount in a hemocytometer and count the number of cells.

8. Subsequently, determine the cell density and use 20,000 cells/cm^2 of the well for seeding. Prepare the desired cell density by diluting the initial cell suspension with the appropriate volume of cell medium.

9. Transfer the cell suspension into collagen precoated 6-well flexible bottomed culture plates in a total volume of 3 mL/well.

10. Allow the cells to grow until confluent at 37 °C for 1 week. Change the culture medium twice per week.

3.2 Differentiation of cEND Cells Prior to Stretch

1. Cells require differentiation before stretch injury. In order to induce differentiation of cells in culture, change the regular culture medium with medium containing 1% serum-stripped fetal calf serum (ssFCS) (*see* **Note 4**).

2. Incubate the cells at 37 °C for 24 h.

Table 1
A guide for setting the degree of stretch injury

Regulator pressure (psi)	Peak pressure (psi)	Degree of injury
15	1.2–1.5	<Low
20–25	1.8–2.0	Low
30–35	2.5–3.0	Moderate
40–50	3.5–4.5	Severe
60	4.8–6.0	>Severe

3.3 Stretch-Induced Injury of cEND Cells

1. Turn on the cell injury controller device. Set the delay to 50 ms and the regulator pressure to 15 psi or an experimentally tested psi depending on the desired severity of injury using the knob. Make sure that the peak pressure is stable by pressing the trigger a couple of times until the gauge registers a stable value of between 3.5 and 4.5 psi. Depending on the pressure applied, the cells can be subjected to low, moderate or severe stretch. Table 1 provides a guide for generating various degrees of stretch injury and set the desired regulator pressure according to the desired injury (*see* Fig. 2).

2. Place the 6-well flexible bottomed culture plate into the tray holder. When using a small plate, use the small adapter plug and the big plug for a large plate. Make sure that the well selector is set to the correct well size before proceeding.

3. Place the adapter plug firmLy over the well. Hold the plug firmLy into place with one hand while the other hand presses the trigger (*see* **Note 5**).

4. Record the peak pressure generated upon pushing the trigger (*see* **Note 6**).

5. Put the plate back immediately into the 37 °C incubator for 15 min before evaluation. Incubation may be varied depending on the experiment (*see* **Note 7**).

3.4 Oxygen–Glucose Deprivation (OGD) of Stretched Cells

1. After allowing the cells to recover for 15 min after stretch injury, wash the cells with PBS. Afterward, replace the cell culture medium with medium lacking D-glucose and 1% serum-stripped fetal calf serum.

2. Next, incubate the cells in a hypoxic work station or incubator with the following conditions: 0.5% O_2, 5% CO_2, saturated humidity atmosphere and 37 °C for 4 h.

3. After 4 h, change the medium with endothelial cell culture medium containing glucose. The cells may be reoxygenated by allowing them to recover in the 37 °C for 20 h, or may be evaluated immediately.

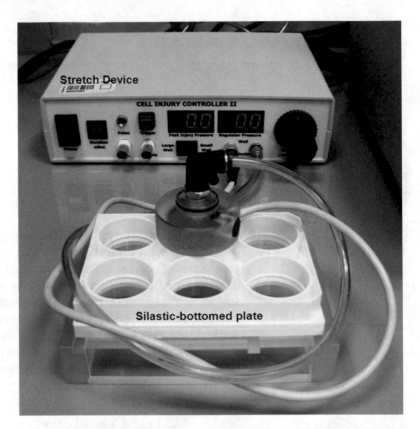

Fig. 2 Stretch injury setup. The 6-well silastic-bottomed plate containing the cells is firmLy placed over the tray. The adapter plug is then placed over the well and the trigger is pushed to stretch the cells

3.5 Live/Dead Staining and Cytotoxicity Assay

1. For assessment of cell viability, apply 100 nm of any viability stain of choice to the cells immediately after stretch and/OGD. Next, view the cells under the microscope (*see* Fig. 3).

2. Alternatively or in addition, after stretch and/or OGD, measure cytotoxicity by assessment of lactate dehydrogenase (LDH) enzyme release. Collect culture medium from treated cells for measurement of LDH released by the cells.

3. Afterward, wash the cells with PBS 2×. Aspirate the medium and add the lysis buffer. Collect the lysed cells and centrifuge at maximum speed for 5 min. Discard the cell pellet and use the supernatant to measure the maximum releasable LDH in the cells.

4. Pipette an appropriate amount of the samples into a 96-wells plate. Add the reaction mixture composed of LDH catalyst and dye solution. Incubate at 37 °C for 30 min (*see* **Note 8**).

5. Finally, add the stop solution to each well. Shake for 10 s. Read the absorption using a microplate reader at 490 or 495 nm.

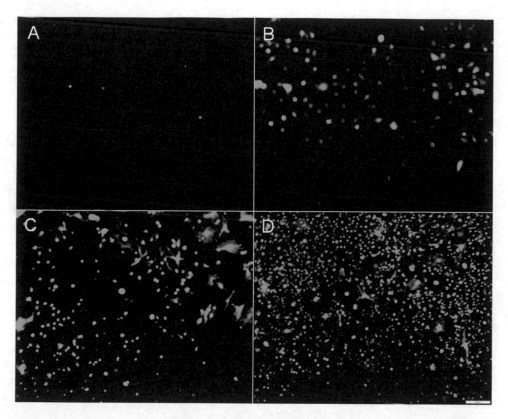

Fig. 3 Fluorescence microscopy of stretched brain microvascular endothelial cells. Cells become permeant to the viability dye when injured. Fluorescence increases with increased degree of injury. (**a**) No stretch (**b**) Low Stretch (**c**) Moderate Stretch (**d**) Severe Stretch. Scale bar: 100 μm

4   Notes

1. cEND cells can be generated by isolation of murine brain capillaries [9, 10]. Other brain endothelial cells may also be used. cEND cells are sensitive to abrupt changes in temperature. Thus, it is recommended that handling of the cells outside the 37 °C incubator be kept to a minimum. If a heat plate is available, it is suggested to keep the cells in contact with the plate set at 37 °C as much as possible at all times during handling. Medium and PBS need to be prewarmed at 37 °C prior to use.

2. Although cEND cells remain relatively adherent to the bottom of the wells during handling, care is still needed with regard to washing the cells. Make sure that the cells will remain intact by pipetting in the PBS through the wall of the well and not on top of the cells. Do the same when aspirating the medium or PBS during washing or medium replacement.

3. When the cells are left to be in trypsin too long, there is a tendency for them not to be able to survive. Therefore, it is good to start with a short trypsinization time of 2 min. If the cells are already detached within this time, one does not have to do it in 5 min.

4. ssFCS can be prepared by adding activated charcoal to heat inactivated fetal calf serum to a final concentration of 1% and Dextran T-70 to a final concentration of 0.1%. The mixture should be incubated for 1 h at room temperature (RT) with agitation in a shaker. Next, the charcoal is pelleted by centrifugation at $12,000 \times g$ for 15 min. The supernatant is then filtered through a vacuum using Whatman filter #4. Afterward, it is filter-sterilized with a 0.2 μM sterile filter. Aliquots are stored at $-20\ °C$ prior to use [13].

5. Ensure that the adapter plug is set firmLy in place and that there is no open space between it and the well as this could result to variations in generated peak pressure from well to well. To achieve uniform peak pressure, the registered pressure should be stable. It is suggested that a cell-free well containing the same amount of medium as those to be used in the experiment be available for calibration. Use this well to generate a stable peak pressure by pushing the trigger several times while the adapter plug is held over the well, before proceeding with the actual experiment.

6. It is not unusual to generate varying peak pressures when performing the stretch injury. Recording the registered peak pressure would aid in the exclusion of wells that received peak pressures which do not fall within the desired range.

7. One should take care that stretch injury be performed with as little time as possible since exposure of the cells to varying temperatures is not good and could consequently affect the results. It is also advised that the cells be put back immediately to the $37\ °C$ incubator after stretch injury.

8. It is very important to protect the samples from light when performing this step. Variations and deviations in the results may be avoided if this is done.

References

1. Morrison B III, Elkin BS, Dolle JP, Yarmush ML (2011) In vitro models of traumatic brain injury. Annu Rev Biomed Eng 13:91–126

2. Morrison B III, Saatman KE, Meaney DF, McIntosh TK (1998) In vitro central nervous system models of mechanically induced trauma a review. J Neurotrauma 15:911–928

3. Ellis EF, Mckinney JS, Willoughby KA, Liang S, Povlishock JT (1995) A new model for rapid stretch-induced injury of cells in culture - characterization of the model using astrocytes. J Neurotrauma 12:325–339

4. Berrout J, Jin M, O'Neil RG (2012) Critical role of TRPP2 and TRPC1 channels in stretch-

induced injury of blood-brain barrier endothelial cells. Brain Res 1436:1–12

5. McKinney JS, Willoughby KA, Liang S, Ellis EF (1996) Stretch-induced injury of cultured neuronal, glial and endothelia cells (Effect of polyethylene glycol-conjugated superoxide dismutase). Stroke 27:934–940

6. Wanner IB, Deik A, Torres M, Rosendahl A, Neary JT, Lemmon VP, Bixby JL (2008) A new in vitro model of the glial scar inhibits axon growth. Glia 56(15):1691–1709

7. Webster GD, Rzigalinski BA, Gabler HC (2008) Development of an improved injury device for neural cell cultures. Biomed Sci Instrum 44:483–488

8. Salvador E, Neuhaus W, Foerster C (2013) Stretch in brain microvascular endothelial cells (cEND) as an in vitro traumatic brain injury model of the blood brain barrier. J Vis Exp 80

9. Burek M, Salvador E, Forster CY (2012) Generation of an immortalized murine brain microvascular endothelial cell line as an in vitro blood brain barrier model. J Vis Exp 66

10. Forster C, Silwedel C, Golenhofen N, Burek M, Kietz S, Mankertz J, Drenckhahn D (2005) Occludin as direct target for glucocorticoid-induced improvement of blood-brain barrier properties in a murine in vitro system. J Physiol Lond 565:475–486

11. Abbruscato TJ, Davis TP (1999) Combination of hypoxia/aglycemia compromises in vitro blood-brain barrier integrity. J Pharmacol Exp Ther 289:668–675

12. Kleinschnitz CBK, Kahles T, Schwarz T, Kraft P, Göbel K, Meuth SG, Burek M, Thum T, Stoll G, Förster C (2011) Glucocorticoid insensitivity at the hypoxic blood–brain barrier can be reversed by inhibition of the proteasome. Stroke 42:1081–1089

13. Dembinski TC, Leung CKH, Shiu RPC (1985) Evidence for a novel pituitary factor that potentiates the mitogenic effect of estrogen in human-breast cancer-cells. Cancer Res 45:3083–3089

Chapter 18

An In Vitro Oxygen–Glucose Deprivation Model for Studying Ischemia–Reperfusion Injury of Neuronal Cells

Myoung-gwi Ryou and Robert T. Mallet

Abstract

Ischemia–reperfusion syndromes of the heart and brain are the leading cause of death and long-term disability worldwide. Development of effective treatments for myocardial infarction, stroke, cardiac arrest and their sequelae requires preclinical models that replicate specific features of ischemia–reperfusion. The complexities of intact animals, including the integrated function of organ systems, autonomic innervation and endocrine factors, often preclude detailed study of specific components of ischemia–reperfusion injury cascades. Ischemia represents the interruption of metabolic fuel and oxygen delivery to support cellular oxidative metabolism; reintroduction of oxygen upon reperfusion of ischemic tissue triggers oxidative stress which initiates the reperfusion injury cascade culminating in injury and death of cells and tissues. Thus, cultured cells subjected to hypoxia, fuel deprivation and reoxygenation replicate the cardinal features of ischemia–reperfusion, while accommodating interventions such as siRNA suppression of specific genes and pharmacological activation or inhibition of signaling cascades that are not feasible in more complex preparations, especially intact animals. This chapter describes an in vitro OGD-reoxygenation cell culture model, an excellent preparation to examine the cellular mechanisms mediating ischemia–reperfusion injury and/or cytoprotection.

Key words Apoptosis, Cell culture, Ischemia–reperfusion (I/R) neurons, Oxygen–glucose deprivation (OGD)

1 Introduction

Ischemia and reperfusion (I/R) injury is the central pathogenic event in many of the leading maladies afflicting the human population, including ischemic stroke [1], coronary artery disease [2], and circulatory arrest [2], and is a frequent comorbidity of invasive cardiothoracic and vascular surgeries. Despite extensive preclinical and clinical efforts, recombinant tissue plasminogen activator (rtPA) remains the only FDA-approved clinical treatment for ischemic stroke, and its effective time window is limited to the first 4.5 h after stroke onset. Because of rtPA's limited treatment window, only 4–5% of stroke patients benefit from rtPA treatment [3]. The failure to identify robust treatments could be due in part to

Binu Tharakan (ed.), *Traumatic and Ischemic Injury: Methods and Protocols*, Methods in Molecular Biology, vol. 1717,
https://doi.org/10.1007/978-1-4939-7526-6_18, © Springer Science+Business Media, LLC 2018

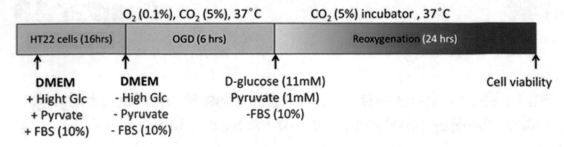

Fig. 1 Proposed OGD-reoxygenation model. *Glc* Glucose, *OGD* oxygen–glucose deprivation, *FBS* fetal bovine serum

limitations of experimental models and protocols, particularly in in vitro preparations. On the other hand, carefully conducted in vitro OGD-reoxygenation experiments will provide fundamental and meaningful information regarding molecular mechanisms of ischemia–reperfusion-induced injury in heart and brain, particularly myocardial infarct, ischemic stroke, and cardiac arrest. Currently, numerous in vitro OGD-reoxygenation models have been developed, but detailed, step-by-step documentation of the experimental protocols oftentimes is not readily available, making recapitulation of the experiments challenging at best. Accordingly, the goal of this chapter is to provide methodological details regarding oxygen-glucose deprivation experiments in cultured cells, a facile in vitro model of ischemia–reperfusion, so as to assist investigators performing in vitro OGD-reoxygenation experiments (*see* Fig. 1) in a commercially available, auto-O_2 regulated hypoxia chamber.

2   Materials

All solutions were prepared with ultrapure water, produced by distilling deionized water to attain a resistance of ≥ 18 MΩ cm at 25 °C. Mouse hippocampus derived neuronal HT22 cells of <20 passages were used in this protocol.

2.1   Cell Culture

1. High glucose (25 mM) Dulbecco's minimal essential medium (DMEM; Gibco, Grand Island, NY, USA) was stored at 4 °C.

2. Antibiotics: Penicillin and streptomycin cocktail (Sigma-Aldrich, St. Louis, MO).

3. Phosphate-buffered solution (PBS): 150 mM NaCl, 2.6 mM KCl, 8.1 mM Na_2HPO_4, and 2.5 mM KH_2PO_4. pH was adjusted to 7.4.

2.2   Oxygen–Glucose Deprivation (OGD) Stress

1. Media: DMEM without glucose or pyruvate: 0.4 mM glycine, 0.4 mM L-Arginine hydrochloride, 0.2 mM L-cysteine 2HCl, 4 mM L-glutamine, 0.2 mM L-isoleucine, 0.8 mM L-leucine,

0.79 mM L-lysine hydrochloride, 0.2 mM L-Methionine, 0.4 mM L-Phenylalanine, 0.4 mM L-Serine, 0.8 mM L-Threonine, 0.08 mM L-Tryptophan, 0.4 mM L-Tyrosine disodium salt dehydrate, 0.8 mM L-Valine, 0.03 mM Choline chloride, 0.01 mM D-Calcium pantothenate, 0.01 mM Folic Acid, 0.03 mM Niacinamide, 0.02 mM Pyridoxine hydrochloride, 0.001 mM Riboflavin, 0.012 mM Thiamine hydrochloride, 0.04 mM i-Inositol, 1.8 mM Calcium Chloride ($CaCl_2$) (anhyd.), 2.48 mM Ferric Nitrate ($Fe(NO_3)_3 \cdot 9H_2O$), 0.8 mM Magnesium Sulfate ($MgSO_4$) (anhyd.), 5.33 mM Potassium Chloride (KCl), Sodium Bicarbonate ($NaHCO_3$), 110.3 mM Sodium Chloride (NaCl), 0.9 mM Sodium Phosphate monobasic ($NaH_2PO_4 \cdot H_2O$), 0.04 mM Phenol Red.

Locke's media: 154 mM NaCl, 5.6 mM KCl, 2.3 mM $CaCl_2$, 1.0 mM $MgCl_2$, 3.6 mM $NaHCO_3$, 5 mM HEPES. pH was adjusted to 7.2 to mimic ischemia-induced acidemia.

2. Hypoxia Chamber: Auto-controlled hypoxia chambers (Bio-Spherix, Lacona, NY, USA) were used in this protocol to impose severe hypoxia (0.1% O_2 atmosphere).

2.3 Reoxygenation

Reoxygenation Solution: 100 mM Na-Pyruvate, 1.1 M d-glucose

2.4 Cell Viability Assay

1. Calcein AM (1 μmol/L), fluorescent microplate reader (Infinite 200 pro, TECAN, San Jose, CA, USA).

   Add 1 mL of Dimethyl Sulfoxide (DMSO) to calcein AM stock (1 mM).

2. Annexin-V.

3. Propidium iodide.

4. 10× binding buffer: HEPES (0.1 M, pH 7.4), NaCl (1.4 M), $CaCl_2$ (25 mM).

5. Flow cytometry (BD LSR II, San Jose, CA, USA).

3 Methods

A variety of cultured cells could be exposed to the OGD-reoxygenation stress. However, depending on cell type, several factors must be considered, including 1) Duration of pre OGD incubation, 2) Seeding cell density, 3) Composition of culture and OGD media, 4) Reoxygenation sequence. The following text describes a protocol designed to study mouse hippocampus-derived neuronal HT22 cells.

3.1 Cell Preparation

HT22 cells are mouse hippocampus derived neuronal cells, which are not oncogenic cells. Attachment to culture plate and stabilization of HT22 cells is relatively fast in comparison to other primary cells, such as astrocytes and neurons. The larger number of attached cells renders these cells more resistant than other cells to the OGD stress, so the duration of seeding prior to OGD is critical. HT22 cells (5×10^3 cells/well in 96-well plate, 1×10^5 cells/wells in 6-well plate) were seeded for 16–20 h before imposing OGD, at which time cells should be approximately 60% confluent. Higher degrees of confluence will minimize the impact of OGD on these cells.

3.2 Oxygen–Glucose Deprivation (OGD)

1. Cells are washed with PBS (pH 7.4) three times (*see* **Note 1**).

2. DMEM without glucose and pyruvate was added to cell culture well (*see* **Note 2**).

   Alternative OGD buffers can be used to place the cells under more extreme OGD stress. An example is glucose-free extracellular Locke's medium (*see* Subheading 2.2, **item 2**).

3. Hypoxia chamber setting: Set the O_2 concentration to 0.1% and duration 6 h (*see* **Notes 3** and **4**).

3.3 Reoxygenation

Immediately after scheduled OGD treatment, process following reoxygenation treatment.

1. 1 μl of 100× pyruvate and dextrose mix is added to 100 μl of OGD media (96-well plate preparation) at the beginning of reoxygenation (*see* **Notes 5** and **6**).

3.4 Evaluation of OGD and Reoxygenation Stress

Calcein AM assay and apoptotic cell deaths are determined with flow cytometry along with Annexin-V and propidium iodide stain (*see* **Note 7**).

3.5 Calcein AM Assay

1. Dilute 1 vol calcein AM stock in 1000 vol PBS.

2. After OGD-reoxygenation, cells are carefully washed with PBS (pH 7.0).

3. Add 100 μl calcein AM and return the plate to the incubator for 10–30 min (longer incubations will affect nearby wells).

4. Dump the calcein AM solution from plate (invert over sink with force) (*see* **Note 8**).

5. Rinse the wells with PBS (add 150 μl PBS and then dump this solution) (*see* **Note 8**).

6. Wrap plate with aluminum foil to prevent light exposure (*see* **Note 9**).

7. Allow the plate to sit at room temperature for 15 min and then read fluorescence.

8. Image could be taken under fluorescent microscope (*see* Fig. 2).

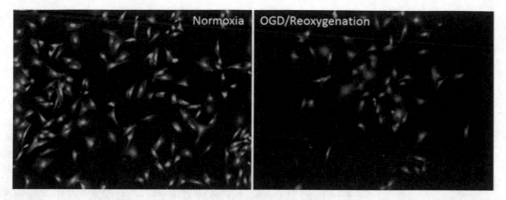

Fig. 2 Cell viability assay image by calcein AM. HT 22 cells were treated with either normoxia (left) or OGD-reoxygenation (right). Normoxic HT22 cells maintain morphological integrity and express more green fluorescent intensity than cells subjected to OGD-reoxygenation

3.6 Flow Cytometry with Annexin-V and Propidium Iodide Stain [4]

1. Wash the cells with PBS ($2\times$).

2. Resuspend cells in $1\times$ binding buffer (1×10^6 cells/mL).

3. Transfer 100 µl of the solution (1×10^5 cells) to a 5 mL culture tube.

4. Add 5 µl of FITC Annexin V.

5. Add 10 µl of propidium iodide (*see* **Note 10**).

6. Gently mix the cells by vortex and incubate for 15 min at room temperature in the dark.

7. Add 400 µl of $1\times$ binding buffer to each tube.

8. Analyze by flow cytometry within 1 h (*see* **Note 11**).

4 Notes

1. In order to ensure consistent basal sample conditions, it is important to minimize the loss of the cells during washing process.

2. 100 µl media/well, 96-well plate; and 1.5 mL media/well, 6-well plate.

3. The chamber's auto-control system permits O_2 concentration to be adjusted between 0.1% and 21% O_2. Depending on tissue and organ, in situ O_2 concentration is about 2–3% [5]. The diffusion of hypoxic air through culture medium slowly occurs which might limit the rate of O_2 adjustment. For reference, in cell preparations contained in 10 cm culture plates with 10 mL of media in auto-controlled hypoxia chambers (BioSpherix, Lacona, NY, USA), it takes about 20–30 min for media O_2 concentration to equilibrate to the chamber O_2 concentration.

Therefore, it is important to minimize the volume of culture media to facilitate more rapid equilibration [6].

4. Duration of OGD conditioning: It is important to determine the appropriate OGD exposure duration before proceeding to your experiment. For experiments to study the protective effect of an agent, choose a duration that causes 40–50% cell death in the absence of the treatment. If cell death is too extreme, e.g., $\geq$80%, the effects of agents will be difficult to detect or quantify in vitro setting. In the case of HT22 cells, 4–6 h exposure to 0.1% O_2 will produce *c.* 50% cell death.

5. The final concentrations of glucose and pyruvate in media were 11 mM and 1 mM, respectively. By adding 100× dextrose and pyruvate to preexisting OGD media at the beginning of reoxygenation, metabolites and free radicals generated during OGD are still able to influence the experimental outcomes. Concentrated glucose and pyruvate will minimize dilution of metabolites and of specific treatments.

6. Reoxygenation mimics the reperfusion phenomenon. Therefore, it is important to retain the metabolites generated during OGD, and the timing of treatment at the beginning of reoxygenation is critical. A typical mistake is to completely remove OGD media at the beginning of reoxygenation and replace it with complete culture media.

7. OGD-reoxygenation stress-induced cellular damage is readily assessed by various assays, such as Calcein AM assay, MTT (3-(4,5-Dimethylthiazol-2-Yl)-2,5-Diphenyltetrazolium Bromide) assay, lactate dehydrogenase release, and measuring apoptotic (or necrotic) markers.

8. Especially after longer OGD and reoxygenation, cells can be easily detached from plate, so careful handling is required.

9. Fluorescent intensity from calcein AM is sensitive to light exposure, and therefore it is important to avoid exposing the preparation to light.

10. The optimal concentration of propidium iodide may vary among cell lines, but 10 μl of a 50 μg/mL stock is most likely the maximum concentration required. Lower PI concentrations may yield optimal results in some experimental systems.

11. Keep the sample on ice without light exposure.

Acknowledgments

This work was supported by UNTHSC intramural research grant RI6148 and grant NS076975 from National Institute of Neurological Disorders and Stroke.

References

1. Ryou MG, Choudhury GR, Li W, Winters A, Yuan F, Liu R, Yang SH (2015) Methylene blue-induced neuronal protective mechanism against hypoxia-reoxygenation stress. Neuroscience 301:193–203

2. Frank A, Bonney M, Bonney S, Weitzel L, Koeppen M, Eckle T (2012) Myocardial ischemia reperfusion injury: from basic science to clinical bedside. Semin Cardiothorac Vasc Anesth 16(3):123–132

3. Green AR (2008) Pharmacological approaches to acute ischaemic stroke: reperfusion certainly, neuroprotection possibly. Br J Pharmacol 153 (Suppl 1):S325–S338

4. Xie L, Li W, Winters A, Yuan F, Jin K, Yang S (2013) Methylene blue induces macroautophagy through 5′ adenosine monophosphate-activated protein kinase pathway to protect neurons from serum deprivation. Front Cell Neurosci 7:56

5. Choi JR, Pingguan-Murphy B, Wan Abas WA, Yong KW, Poon CT, Noor Azmi MA, Omar SZ, Chua KH, Xu F, Wan Safwani WK (2015) In situ normoxia enhances survival and proliferation rate of human adipose tissue-derived stromal cells without increasing the risk of tumourigenesis. PLoS One 10(1): e0115034

6. Yang C, Jiang L, Zhang H, Shimoda LA, DeBerardinis RJ, Semenza GL (2014) Analysis of hypoxia-induced metabolic reprogramming. Methods Enzymol 542:425–455

Chapter 19

Measurement of Microvascular Endothelial Barrier Dysfunction and Hyperpermeability In Vitro

Bobby Darnell Robinson, Chinchusha Anasooya Shaji, Angela Lomas, and Binu Tharakan

Abstract

Loss of microvascular endothelial barrier integrity leads to vascular hyperpermeability and vasogenic edema in a variety of disease processes including trauma, ischemia and sepsis. Understanding these principles gives valuable information on pathophysiology and therapeutic drug development. While animal models of traumatic and ischemic injuries are useful to understand vascular dysfunctions associated with such injuries, in vitro barrier integrity assays are reliable and helpful adjuncts to understand the cellular and molecular changes and signaling mechanisms that regulate barrier function. We describe here the endothelial monolayer permeability assay and transendothelial electrical resistance (TEER) measurement as in vitro methods to test changes in microvascular integrity and permeability. These in vitro assays are based on either the measurement of electrical resistance of the monolayer or the quantitative evaluation of fluorescently tagged molecules (e.g., FITC-dextran) that pass through the monolayer when there is damage or breakdown.

Key words Microvascular permeability, Endothelium, Monolayer permeability, TEER

1 Introduction

The endothelium is metabolically active and plays an important role in physiological processes such as the control of vasomotor tone, the trafficking of leukocytes between blood and underlying tissue, angiogenesis, and both innate and adaptive immunity [1]. Endothelial cells provide a nonthrombogenic monolayer surface that lines the lumen of blood vessels and functions as a cellular interface between blood and tissue. Epithelial cells line and provide a protective layer for the cavities and lumens of the body. Epithelial and endothelial cells are connected to each other via intercellular junctions that differ in their morphological appearance, composition, and function. The tight junction is the intercellular junction that regulates diffusion and allows both of these cell layers to form selectively permeable cellular barriers that separate apical (luminal) and basolateral (abluminal) sides in the body, thereby controlling

Binu Tharakan (ed.), *Traumatic and Ischemic Injury: Methods and Protocols*, Methods in Molecular Biology, vol. 1717, https://doi.org/10.1007/978-1-4939-7526-6_19, © Springer Science+Business Media, LLC 2018

the transport processes to maintain homeostasis. Barrier integrity is vital for the physiological activities of the tissue [2].

The function of exchange vessels is to allow the unhindered transfer of dissolved gases, ions, and solutes across the vessel wall. The vast majority of these substances are low in molecular weight and higher in concentration in the plasma than in the interstitium; thus, passive diffusion is the chief transport mode for these solutes [3]. The transport of macromolecules larger than 3 nm, such as albumin, IgG, and other macromolecules, occurs transcellularly through transcytosis or vesicular transport. Molecules smaller than 3 nm, such as glucose, water, and ions, can pass through paracellular pathways [4]. Vasoactive substances and mechanical stress triggered by hemodynamic forces, such as mechanical stretch and shear stress, stimulate endothelial cell signaling [5].

Epithelial and endothelial disruption leads to many of the derangements seen in trauma. To maintain normal brain function, the neural environment must be preserved within a narrow homeostatic range; this requires a tight regulation of transportation of cells, molecules and ions between the blood and the brain. Such tight regulation is maintained by a unique anatomical and physiological barrier, formed collectively in the central nervous system (CNS). Blood–brain barrier (BBB) disruption leads to cerebral edema causing elevated intracranial pressure (ICP), decreased cerebral blood flow, poor tissue oxygenation, brain herniation, and induction of apoptotic cell death [6]. These factors have increased patient morbidity and mortality. Endothelial permeability in the lungs causes pulmonary edema and causes respiratory distress; this damage not only causes exudate formation but also attenuates its clearance [7]. Damage to the tight junctions of the gastrointestinal system may precede and promote translocation of bacteria (migration of microbes or their products into mesenteric lymph nodes), endotoxins (such as lipopolysaccharides), and pathogen-associated molecular patterns (PAMPs) into the portal venous system and extraintestinal sites [8].

Endothelial cell dysfunction and breakdown can be studied in various traumatic situations including burn, hemorrhage, sepsis, and traumatic brain injury. The two in vitro techniques to measure that will be discussed in this chapter are: monolayer permeability assay and transepithelial/transendothelial electrical resistance (TEER) measurement. The monolayer permeability assay creates a functional endothelium with a single layer of endothelial cells and studies the permeability of a fluorescent solute as it passes from the luminal to abluminal compartment. Between 4 and 6 data points (wells) is sufficient to find statistical difference. Sample readings can be converted with the use of a standard curve to albumin concentration. These concentrations were then used in the following equation to determine the permeability coefficient of albumin (P_a), $P_a = [A]/t \times 1/A \times V/[L]$, where $[A]$ = abluminal concentration, t = time in s, A = area of membrane in cm^2, V = volume of abluminal chamber, and [L] = luminal concentration [9].

The electrical resistance of a cellular monolayer, measured in ohms, is a quantitative measure of the barrier integrity. The classical setup for measurement of TEER consists of a cellular monolayer cultured on a semipermeable filter insert that defines a partition for apical (or upper) and basolateral (or lower) compartments. For electrical measurements, two electrodes are used, with one electrode placed in the upper compartment and the other in the lower compartment, and the electrodes are separated by the cellular monolayer. The ohmic resistance is calculated based on Ohm's law as the ratio of the voltage and current. The measurement procedure includes measuring the blank resistance (R_{BLANK}) of the semipermeable membrane only (without cells) and measuring the resistance across the cell layer on the semipermeable membrane (R_{TOTAL}): R_{TOTAL} includes the ohmic resistance of the cell layer R_{TEER}, the cell culture medium R_M, the semipermeable membrane insert R_I, and the electrode medium interface R_{EMI}. The cell-specific resistance (R_{TISSUE}), in units of Ω, can be obtained as: R_{TISSUE} (Ω) = R_{TOTAL} − R_{BLANK}. Then reported as: TEER$_{REPORTED}$ = R_{TISSUE} (Ω) × Membrane (M)$_{AREA}$ (cm^2). Resistance is inversely related to permeability.

2 Materials

1. Human Brain Microvascular Endothelial Cells (HBMECs), Rat Brain Microvascular Endothelial Cells (RBMECs) and RBMEC Medium.

2. Opti-MEM (MEM)/reduced serum medium and Dulbecco's modified Eagle's medium (DMEM; with high glucose, HEPES, no phenol red).

3. Phosphate buffered saline (PBS).

4. Fluorescein isothiocyanate (FITC)-dextran.

5. Permeability assay performed with Costar Transwell membranes.

6. Fluorometry measured at 485/520 nm (Excitation/Emission) using Fluoroskan Ascent™ FL Microplate Fluorometer and Luminometer.

7. Epithelial Volt/Ohm Meter (EVOM2) together with STX2 electrode purchased from World Precision Instruments.

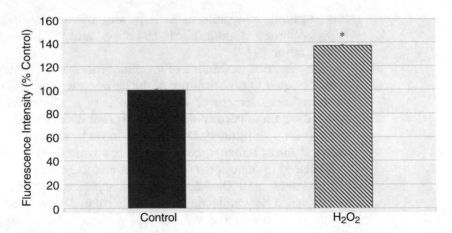

Fig. 1 Monolayer permeability study using human brain microvascular endothelial cells (HBMECs). Control vs. Treatment with 200 µM H_2O_2 for 2 h. Presented as percent control. H_2O_2 increased permeability significantly. *$p < 0.05$

3 Methods

3.1 Monolayer Permeability Assay

1. Grow RBMECs/HBMEC attached to 100 mm plates to confluency.

2. Put 100 µl fibronectin in to chamber; allow it for 30–60 min in 37 °C incubator for cell attachment.

3. RBMECs were seeded at a density of 10^5 cells/cm^2 on the fibronectin-coated Costar Transwell membranes

4. Add 600 µl media to unused lower chambers, now transfer inserts to lower chamber.

5. Add 100 µl of cells to each well.

6. Incubate the cells at 37 °C until confluent, typically 3–5 days depending on the cell type

7. Aspirate the old media and add 600 µl of phenol red free media such as or Opti-Minimal Essential Medium (MEM) for 1 h, both in upper (100 µl) and lower chamber (600 µl)

8. Treat with required inhibitor for 1–3 h (depending on inducer)

9. Add inducer for 1–3 h to inducer inserts and inducer + inhibitor inserts only (*see* **Note 1**)

10. Expose to 10 µL of 5 µg/mL FITC-dextran for 30 min

11. Collect 100 µl from lower chamber and read in opaque black 96 well plate @485 and 520 nm in fluorimeter (Fig. 1).

3.2 TEER

1. Place apparatus on a heating pad at 37–40 °C.

2. Wash electrode and endohm chambers twice with deionized (DI) water.

3. Connect all wires but do not turn on.

4. Aliquot DMEM media that you used for the monolayer study to use for your TEER readings.

5. Soak the upper and lower electrode in this media for 10 min prior to taking the reading, with the EVOM2 (submerge top electrode with media).

6. After priming the electrodes (soaking them in the desired media) remove the media from the lower chamber of the endohm.

7. Add 1 mL of the desired media into the lower chamber of the endohm (on heating pad).

8. Remove the desired monolayer plate from the incubator and place it next to endohm.

9. Take the cell monolayer inserts from the tray and place into the lower chamber of the endohm (membrane should NOT sit on the bottom electrode. There needs to be space between bottom of monolayer membrane and the bottom electrode so that there is a free flow of current) (Fig. 2).

10. Place the top electrode on the endohm. Ensure the top electrode is submerged in the media of the monolayer chamber (*see* **Notes 2** and **3**).

11. View the level of media in the bottom chamber in relation to the level of media in the monolayer with the top electrode in it. Try to get the level of media in the lower chamber and in the monolayer chamber to be more or less equal (*see* **Note 4**).

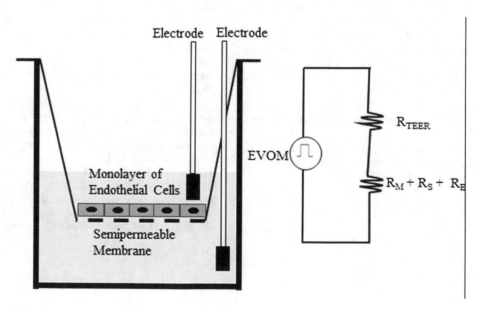

Fig. 2 Classic TEER setup with chopstick electrodes

12. When media levels are equal, wait for the readings on the EVOM2 to stabilize (same reading for 3–6 s) then record.

13. The final unit area resistance (Ω cm^2) was calculated by multiplying the sample resistance by the effective area of the membrane (*see* **Note 5**).

4 Notes

1. Hydrogen peroxide, bradykinin, histamine, or proinflammatory cytokines such as IL-1β are all well-tested inducers of permeability.

2. If the top electrode is out of the media and in the air, or if there are bubbles on the top electrode, then there may be artificially high readings (typically >12,000) on the EVOM2.

3. If the electrodes start to give fluctuating readings, the chloride tips might need to be rechloritized. Soak the electrodes in 5% bleach for 10–15 min (purple black layer forms on electrodes)

4. In standardization attempts, typically 180 µL in upper monolayer chamber with 1 mL in lower chamber gives equal levels, but this may have to be modified based on changes in media and cell type.

5. Multiply by 0.33 cm^2 for Costar Transwell inserts.

References

1. Zielińska KA, Van Moortel L, Opdenakker G, De Bosscher K, Van den Steen PE (2016) Endothelial response to glucocorticoids in inflammatory diseases. Front Immunol 7:592

2. Srinivasan B, Kolli AR, Esch MB, Abaci HE, Shuler ML, Hickman JJ (2015) TEER measurement techniques for in vitro barrier model systems. J Lab Autom 20:107–126

3. Mehta D, Malik AB (2006) Signaling mechanisms regulating endothelial permeability. Physiol Rev 86:279–367

4. Sukriti S, Tauseef M, Yazbeck P, Mehta D (2014) Mechanisms regulating endothelial permeability. Pulm Circ 4:535–551

5. Hirase T, Node K (2012) Endothelial dysfunction as a cellular mechanism for vascular failure. Am J Physiol Heart Circ Physiol 302: H499–H505

6. Logsdon AF, Lucke-Wold BP, Turner RC, Huber JD, Rosen CL, Simpkins JW (2015) Role of microvascular disruption in brain damage from traumatic brain injury. Compr Physiol 5:1147–1160

7. Wittekindt OH (2017) Tight junctions in pulmonary epithelia during lung inflammation. Pflugers Arch 469:135–147

8. Aguirre Valadez JM, Rivera-Espinosa L, Méndez-Guerrero O, Chávez-Pacheco JL, García Juárez I, Torre A (2016) Intestinal permeability in a patient with liver cirrhosis. Ther Clin Risk Manag 12:1729–1748

9. Tinsley JH, MH W, Ma W, Taulman AC, Yuan SY (1999) Activated neutrophils induce hyperpermeability and phosphorylation of adherens junction proteins in coronary venular endothelial cells. J Biol Chem 274:24930–24934

INDEX

Binu Tharakan (ed.), *Traumatic and Ischemic Injury: Methods and Protocols*, Methods in Molecular Biology, vol. 1717,
https://doi.org/10.1007/978-1-4939-7526-6, © Springer Science+Business Media, LLC 2018

Printed in the United States
By Bookmasters